Nanoemulsions - Properties, Fabrications and Applications

Edited by Kai Seng Koh and Voon Loong Wong

Published in London, United Kingdom

IntechOpen

Supporting open minds since 2005

Nanoemulsions - Properties, Fabrications and Applications
http://dx.doi.org/10.5772/intechopen.78812
Edited by Kai Seng Koh and Voon Loong Wong

Contributors
Alejandro Luis Vega-Jiménez, América Vázquez-Olmos, Enrique Acosta-Gío, Marco Antonio Alvarez-Pérez, Lebogang Katata-Seru, Bathabile Ramalapa, Lesego Tshweu, Chaoming Wang, Xinran Zhang, Wenbing Jia, Wei Wu, Louis Chow, Praveen Kumar Gupta, Nividha Bhandari, Hardik N Shah, Vartika Khanchandani, Keerthana R, Vidhyavathy Nagarajan, Lingayya Hiremath, Kaustav Bhattacharjee, Margarita Sanchez-Dominguez, Maira B. Moreno-Trejo, Angela Suarez-Jacobo, Arturo A. Rodríguez-Rodríguez, Kai Seng Koh, Voon Loong Wong

First published in London, United Kingdom, 2019 by IntechOpen
IntechOpen is the global imprint of INTECHOPEN LIMITED, registered in England and Wales, registration number: 11086078, The Shard, 25th floor, 32 London Bridge Street
London, SE19SG – United Kingdom
Printed in Croatia

British Library Cataloguing-in-Publication Data
A catalogue record for this book is available from the British Library

Additional hard and PDF copies can be obtained from orders@intechopen.com

Nanoemulsions - Properties, Fabrications and Applications
Edited by Kai Seng Koh and Voon Loong Wong
p. cm.
Print ISBN 978-1-78984-175-6
Online ISBN 978-1-78984-176-3
eBook (PDF) ISBN 978-1-83881-939-2

For EU product safety concerns:
IN TECH d.o.o., Prolaz Marije Krucifikse Kozulić 3, 51000 Rijeka, Croatia, info@intechopen.com or visit our website at intechopen.com.

Meet the editors

Kai Seng, KOH, is an assistant professor in chemical engineering at Heriot-Watt University, Malaysia. He has a broad range of academic as well as industrial interests, including microscale fluid dynamics, multiphase flow, microelectromechanical device packaging of design and fabrication, energetic material synthesis and characterization, and life science applications using the technology of droplet microfluidics. Dr. Koh currently resides in Kuala Lumpur, Malaysia, with his family and travels to different countries for research inspiration.

Voon-Loong Wong completed his PhD studies at the University of Nottingham in 2015. Currently, he is an assistant professor of chemical engineering at Heriot-Watt University Putrajaya, Malaysia Campus. His research mainly focuses on the formation of microdroplets with non-Newtonian characteristics in microreactors for rheological applications in drug delivery systems. Among his research interests, he focuses on computational fluid dynamics analysis of the droplet interface evolution with a level set approach using COMSOL Multiphysics. He is also actively involved in a range of collaborative projects that relate to separation processes, wastewater treatment, microfluidics, drug delivery systems, and biomedical applications using microfluidics technology.

Contents

Preface

"Fluidics" originally described pneumatic and hydraulic control systems, where fluids were employed (instead of electric currents) for signal transfer and processing. Fluidics then broadened and now comprises the technique of handling fluid flows from the macroscale down to the nanoscale. Nanofluidics is a multidisciplinary field that involves engineering, physics, chemistry, biology/biotechnology, and nanotechnology and was first introduced back in the 1990s. As new as this field may sound, it has gained much attention since its debut as a rising area of interest in nanotechnology. This textbook is written primarily for mature undergraduates in engineering and physics. However, it should be of interest to first-year graduate students and professionals in industry as well.

This book's carefully arranged seven chapters cover fundamental nanofluidics, especially in the area of nanoemulsions, including fundamental knowledge of properties and fabrications to applications in the field. Furthermore, this seven-chapter exposure should provide students with sufficient background for advanced studies in these fascinating and very future-oriented engineering areas, as well as for expanded job opportunities. Pedagogical elements include a 50/50 physics/mathematics approach when introducing new material, illustrating concepts, showing graphical/tabulated results as well as links to flow visualizations, and, very importantly, providing professional problem solution steps.

The ultimate goal of this book is for readers to be able to solve traditional and modern fluidics problems independently, provide physical insight, and suggest (say, via a course project) system design improvements. This text relies on numerous open-source material as well as contributions provided by research associates from across the globe. The standard of the content is of a very high standard.

For critical comments, constructive suggestions, and tutorial material, please contact the author via k.koh@hw.ac.uk.

Kai Seng Koh and Voon Loong Wong
Heriot-Watt University Malaysia,
Putrajaya, Malaysia

Section 1

Introduction

Chapter 1

Introductory Chapter: From Microemulsions to Nanoemulsions

Koh Kai Seng and Wong Voon Loong

1. Introduction

In the past two decades, there has been much attention within food and beverages, pharmaceutical, biomedical, special chemicals and other industries in using colloidal as main media for process encapsulation, protection and delivery of various active components for different purposes [1–3].

This dispersion normally exists as a suspension of small particles within a liquid medium. Conventionally, particles at >1000 nm are studied due to its rising trend of research interest as a result of introduction of microfluidics [4]. Microfluidics is a concept that is defined as a branch of fluid mechanics that focuses on the understanding, design, fabrication and operation systems that convey liquids and gases inside an enclosed channel with two of the three geometry length scales are in the order of microns (10^{-6}). The reduction in dimension had magnified the effect of some uncommon macroscale liquid properties such as surface tension, capillary effect and material hydrophilicity/hydrophobicity. The first ventures in microfluidic started in the early 1950s when dispersion methods of nano- (10^{-9}) and pico- (10^{-12}) litre of fluids were developed, which served as the foundation of modern day Inkjet technology [5, 6]. As microfluidics' continuous development advanced for the past 70 years, multiple cross disciplines with intersections of engineering, physics, chemistry, biology, nanotechnology and biotechnology had been developed and commercialised. In 2003, Forbes magazine named microfluidics technology as one of the most important inventions that can affect the future of humanity [7]. Meanwhile, microfluidics hold some of the key advantages that include the low manufacturing costs, economic use and disposal, shorter time of analysis, minimal consumption of reagents and samples, minimal production of potentially harmful by-products, enhancement of separation efficiency, enhancement of portability for point-of-care testing, high surface to volume ratio and small laboratory footprint [8, 9].

On the other hand, there are some colloidal applications that desire very much smaller particles (<100 nm) since they have advantages over microscale colloids, such as better stability to particles aggregation and gravitational separation [2], weak light scattering [10–12], and have novel physical properties (i.e., high viscosity and gel-like behaviour) [2, 13]. In conjunction with the new rise of nanotechnology research trend, a decent amount of research has been devoted to fabrication, characterization and application of colloidal dispersion that contains of nanometre-sized particles as delivery systems. As the research duration is merely less than two decades, much knowledge gap in this research field remains to be filled.

This chapter therefore emphasizes the most commonly used terms in this field of study, namely microemulsions (>1000 nm) and nanoemulsions (<100 nm). Using oil-in-water (O/W) system, which is widely used as delivery systems in a range of different industries as well as many literatures, this chapter will outline the similarities and differences between microemulsions and nanoemulsions, to articulate the scientific terminology used to describe each of the terms and also to look into research progress of both microemulsions and nanoemulsions as a whole.

2. Terminology

In this section, some fundamental concepts on similarity as well as difference for both microemulsions and nanoemulsions will be provided, respectively, in terms of its physical and chemical properties, stability and thermodynamic difference as well as the current fabrication methods. These terminologies are meant to clarify the common confusions found in most literatures.

2.1 Physical and chemical properties

2.1.1 Microemulsions

The definition of microemulsions is commonly described as a stable liquid droplet in microscale (r < 100 μm) formed by mixing of two immiscible phases, i.e., oil and water in the presence of surfactant. The emulsion's structure, size, composition and surface behaviour can be manipulated by different fabrication methods. They may form one or multiple phases (with different shapes) that are in equilibrium with one another [14]. As for analysis, common analytical equipment such as light-inverted microscopy, scanning electron microscopy, X-ray powder diffraction, electrical conductivity and rheology are usually chosen as tools for the study focus. Therefore, it is advisably to classify microemulsion (O/W) as a thermodynamically stable colloidal dispersion. The structure hinders the unfavourable contact area between non-polar groups and water, which favour the thermodynamics of the colloidal dispersion system. For instance, the surfactant molecules in an oil-in water (O/W) microemulsions has the setup of non-polar tails that associate with each other forming a hydrophobic core. The hydrophilic head groups extrude onto the surrounding aqueous phase of the microemulsion. Meanwhile, the hydrophobic oil molecules may assimilate into the interior of a micelle as a separate core or serves as a barrier between the surfactant tails as shown in **Figure 1**. If oil molecules have the same polar groups, they may then be integrated into the micelle in such a way that it creates a visible distance into the water. This behaviour is particularly important in pharmaceutical industry as it serves as a fundamental core structure for self-microemulsifying drug delivery system. It extends the knowledge for drug delivery system design in terms of the optimum composition of the initial system and the optimum method to dilute the surface of the microemulsion.

2.1.1 Nanoemulsions

Nanoemulsions are considered to be a classic liquid emulsion formation from two immiscible liquids that is thermodynamically unstable. Theoretically, this small spherical droplet (r < 100 nm) could be formed using oil (as continuous phase) and water (as dispersed phase) without addition of a surfactant. However, this system will be highly unstable; thus, most nanoemulsions would require the assistance of surfactant (often it is more than one type of surfactants used) to facilitate its droplet formation. The fundamental component formation of a nanoemulsion is very similar to those found in a microemulsion as

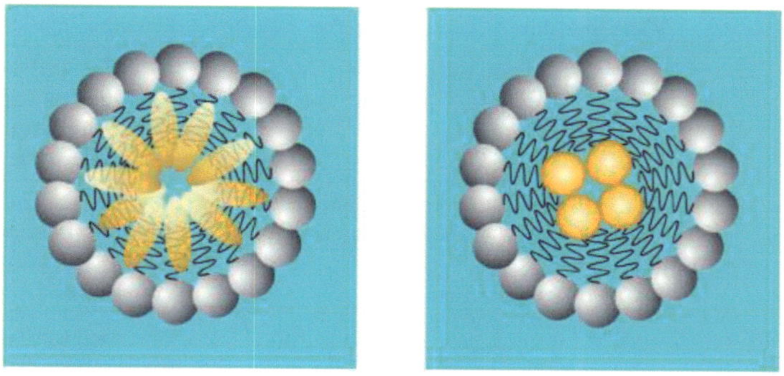

(a) (b)

Figure 1.
Oil-in-water microemulsions: (a) oil molecules assimilate between the surfactant tails; and (b) oil molecules incorporated as a hydrophobic core. Reprinted with permission from Ref. [3].

mentioned above. The only distinctive difference that separates a nanoemulsion from a microemulsion is their thermodynamic stability, i.e., a microemulsion is thermodynamically stable while a nanoemulsion is not. As nanoemulsion only has a short development history of ~25 years [15], there are much area of interest that are waiting to be explored with most of the current understanding for Nanaoemulsions is adapted based on Microemulsions understanding. The summary of similarities and differences between microemulsions and nanoemulsions can be seen in **Table 1**.

The chapter so far outlines the similarity and differences between nanoemulsions and microemulsions based on their general physical and chemical properties, as well as other characteristics. In this section, an attempt to propose practical methods to distinguish nanoemulsions from microemulsions will be made. The key concept to make this comparison is based on thermodynamic point of view as shown in **Figure 2** below. As the terms used for microemulsions and nanoemulsions remain confusing in most literatures, the following two factors can be used as a guideline to distinguish nanoemulsions from microemulsions:

2.1.1.1 Long-term storage

Nanoemulsions are not thermodynamically stable, whereas microemulsions are. This simply means microemulsions will not undergo deformation at an infinite period so long the storage condition remains constant. For nanoemulsions, degradation of structure is noticeable due to Ostwald ripening, flocculation, coalescence and gravitational separation. These result in particle size distribution change, physical properties change as well as chemical properties change. Practically, it can be difficult to just make a justification to distinguish nanoemulsions from microemulsion merely based on long-term storage, since microemulsions commonly suffer from chemical degradation and microbial contamination during the storage period.

2.1.1.2 Particle size

As mentioned above, microemulsions have tendency to form single narrow size distribution, ought to its maturity in terms of fabrication methods. Adding to its thermodynamic stability, the size of a microemulsion will not undergo changes once it is formed by a specific approach. This is different from nanoemulsions where they tend to have multiple peaks in its size distribution. This

Descriptions	Microemulsion	Nanoemulsion	Remarks
Size	1 nm (1×10^{-9} m) to 100 nm (1×10^{-6} m)		The difference between microemulsions and nanoemulsions cannot be distinguished merely from size
Particle size distribution	Tend to have single narrow peak for size distribution	Have multiple peaks for size distribution	Normally, the combination of micro- and nanoemulsions exists in multiple peaks colloid distribution
Water-in-oil/ oil-in-water formation	Possible		Both microemulsion and nanoemulsion can form water-in-oil and oil-in-water emulsion type
Composition	An oil phase, an aqueous phase, a surfactant and possibly a co-surfactant [1]		A greater surfactant-to-oil ratio is required to prepare a microemulsion than a nanoemulsion due to size difference
Optical properties	Opaque or semi-transparent [3]	Transparent when the size is ≤30 nm [12]	
Thermodynamic stability	Kinetically stable for indefinitely	Unstable system that will break down over time [3]	Microemulsions will not undergo breakdown provided the storing condition remains unchanged Nanoemulsions will breakdown and revert back to separated phase, subject to energy barrier between nanoemulsions and separated phase (Gibbs free energy, ΔG)
Gravitational stability	Not reported	Colloidal dispersion to gravitational separation (creaming/ sedimentation) increases significantly for particle size ≤90 nm due to dominance of Brownian motion [3]	Limited work has been to validate this area of research
Particle structure	Spherical or non-spherical, subject to fabrication method used [14]	Spherical	The governing equation is used to determine particle size, thus the structure Laplace pressure, $\left(\Delta P_L = \frac{2\gamma}{r}\right)$. The small radius of nanoemulsions has relative large Laplace pressure, thus sphere sharp has lowest interfacial area
Fabrication methods	Principally, it can form spontaneously due to its thermodynamic stability. In practical, external force will be exerted to emulsion formation	External energy must be applied to overcome positive free Gibbs energy to form emulsion. Common fabrication approaches are high energy, low energy, and phase inversion [3, 16]	Difficult to distinguish based on fabrication methods

Table 1.
Summary of microemulsion and nanoemulsion common facts and common differences.

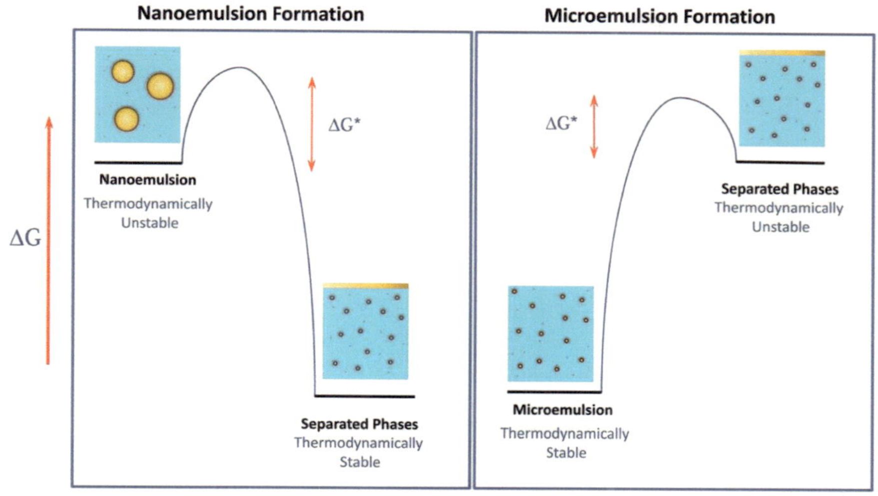

Figure 2.
Schematic illustration between microemulsions and nanoemulsions, with its separated phase state, respectively. Microemulsions have a relative lower free Gibbs energy than the phase-separated state; therefore, it is unlikely to break down even after a long storage period provided the storage condition remains unchanged. Meanwhile, nanoemulsions have a higher free Gibbs energy, leading to breakdown and revert back to its original separated state despite the assistance of surfactants. Reprinted with permission from Ref. [3].

is not surprising due to its unstable thermodynamic as discussed previously. Therefore, if an emulsion system has multiple peaks formation in its size distribution, it can possibly be considered as a system that has both microemulsions and nanoemulsions. However, the noticeable approach to be used in this scenario is to undergo proper physical characterization such as Zetasizer or using scanning electron microscope for proper size measurement, as well as storing it for a period of time, i.e., 1 year to observe the change of emulsion structure. These approaches might provide other significant evidence to validate the type of emulsion formation.

3. Conclusion

This chapter is a timely content of nanoemulsion development as this research field has shown significant publication output for the past decade due to growing interest as a result of nanotechnology development as seen in **Figure 3** using Web of Knowledge (Thomson Reuters) online search engine [15]. Presently, microemulsion-related publication is dominating in terms of publications output per year. This is in line with the history of microemulsion for the past seven decades which covers a comprehensive research range from science (i.e., chemistry, biology, physics, etc.) to engineering (chemical, biomedical, environmental, etc.). On the contrary, nanoemulsions received much lesser focus until the mid-1990s. It has relative incomplete fundamental knowledge understanding in terms of properties, characterizations and fabrication methods, and it also has limited applications presently. Therefore, this book is important in attempting to mend the knowledge gap and form the trend. As nanotechnology matures gradually, the outlook of nanoemulsion looks bright and it aims to stretch its capability to apply across different fields that will benefit mankind.

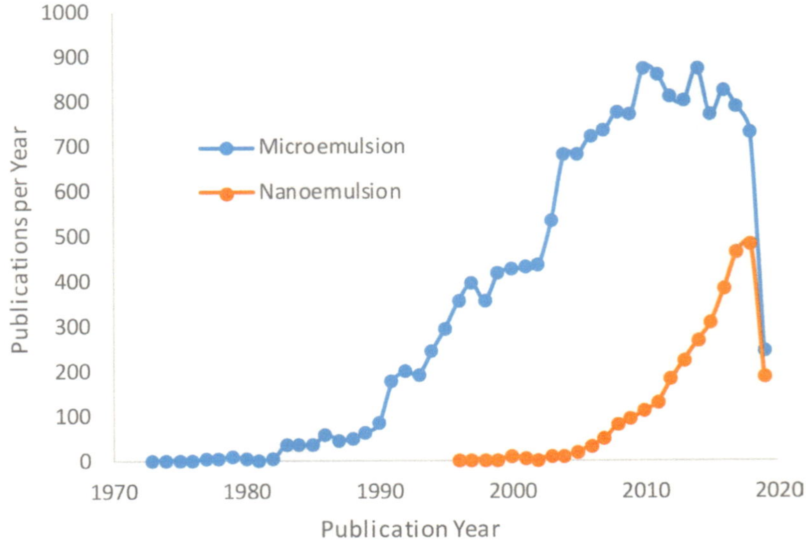

Figure 3.
Comparison between 'microemulsion' and 'nanoemulsion' in terms of publication per year using Web of Knowledge online search engine in May 2019.

Author details

Koh Kai Seng* and Wong Voon Loong
Heriot-Watt University Campus, Malaysia

*Address all correspondence to: k.koh@hw.ac.uk

References

[1] Huang QR, Yu HL, Ru QM. Bioavailability and delivery of nutraceuticals using nanotechnology. Journal of Food Science. 2010;**75**:50-57

[2] Tadros T, Izquierdo R, Esquena J, Solans C. Formation and stability of nano-emulsions. Advances in Colloid and Interface Science. 2004;**108-109**:303-318

[3] McClements DJ. Nanoemulsions versus microemulsions: Terminology, differences, and similarities. Soft Matter. 2012;**8**:1719

[4] Garstecki P, Fuerstman MJ, Stone HA, Whitesides GM. Formation of droplets and bubbles in microfluidic systems. Lab on a Chip. 2006;**6**:163-181

[5] Mark D, Haeberle S, Roth G, von Stetten F, Zengerle R. Microfluidic lab-on-a-chip platforms: Requirements, characteristics and applications. Chemical Society Reviews. 2010;**39**:1153-1182

[6] Berthier J. Micro-Drops and Digital Microfluidics. 2nd ed. Waltham: William Andrew; 2013

[7] Biowww.net. Is Microfluidic Chip Useful? [Online]. 2009. Available from: http://biowww.net/know/1/1573.html [Accessed: 26 November, 2018]

[8] Manz A, Graber N, Widmer HM. Miniaturized total chemical analysis systems: A novel concept for chemical sensing. Sensors and Actuators B: Chemical. 1990;**1**:244-248

[9] Li D. Microfluidics lab-on-a-chip devices for biomedical application. In: Kakaç S, Kosoy B, Li D, Pramuanjaroenkij A, editors. Microfluidics Based Microsystems Fundamentals and Applications. Netherland: Springer; 2010. pp. 377-397

[10] Mason TG, Wilking JN, Meleson K, Chang CB, Graves SM. Nanoemulsions: Formations, structure, and physical properties. Journal of Physics: Condensed Matter. 2006;**18**:635-666

[11] Velikov KP, Pelan E. Colloidal delivery systems for micronutrients and nutraceuticals. Soft Matter. 2008;**4**:1964-1980

[12] Wooster TJ, Golding M, Sanguansri P. Impact of oil type on nanoemulsion formation and ostwald ripening stability. Langmuir. 2008;**224**:12758-12765

[13] Sonneville-Aubrun O, Simonnet JT, L'Alloret F. Nanoemulsions: A new vehicle for skincare products. Advances in Colloid and Interface Science. 2004;**108-109**:145-149

[14] Jonsson B, Lindman B, Holmberg K, Kronberg B. Surfactants & Polymers in Aqueous Solutions. Chichester, U.K: John Wiley & Sons; 1998

[15] Web of Science. 2019. Available from: http://apps.webofknowledge.com/WOS_GeneralSearch_input.do?product=WOS&search_mode=GeneralSearch&SID=F55DgjTcVD6DeepRAnW&preferencesSaved= [Accessed: 28 May, 2019]

[16] McClements DJ. Edible nanoemulsions: Fabrication, properties, and functional performance. Soft Matter. 2011;7:2297-2316

Section 2

Properties

Chapter 2

In vitro Antimicrobial Activity Evaluation of Metal Oxide Nanoparticles

Alejandro L. Vega-Jiménez, América R. Vázquez-Olmos, Enrique Acosta-Gío and Marco Antonio Álvarez-Pérez

Abstract

In recent years, infectious diseases, specifically those that are caused by pathogens, have seen a dramatic proliferation due to resistance to multiple antibiotics, opening the colony by opportunistic pathogens. Nanotechnology and tissue engineering have been applied in the development of new antimicrobial therapies, capable of fighting opportunistic infections. In the medical field, research on antimicrobial properties of metal oxide nanoparticles have emerged to find new antimicrobial agents as an alternative against resistant bacteria. The metal oxides, particularly those formed by transition metals are compounds with electronic properties, and most magnetic phenomena involve this type of oxides. Nanoparticles-based metal oxide properties such as shape, size, roughness, zeta potential and their large surface area, make oxides ideal candidates to interact with bacteria and able to have an antimicrobial effectiveness. The aim of this chapter is to offer an updated panorama about the relationships between the use of metal oxide nanoparticles in the medical field, with an emphasis on their role as antimicrobial agents and the properties that influence their antimicrobial response. In addition, the mechanism of nano-antimicrobial action is described and the importance of using in vitro test methods, adopted by leading international regulatory agencies, that can be used to determine the antimicrobial activity of the metal oxide nanoparticles.

Keywords: nanoparticles, metal oxide, antimicrobial, *in vitro*, methods, mechanism of action, nano-antimicrobial

1. Introduction

Infectious diseases are one of the main causes of morbidity and mortality in the world, so there is the need for research on antimicrobial agents. According to the World Health Organization (WHO), resistance to antimicrobials endangers the effectiveness of treatments for an increasing series of infections by bacteria and fungi [1]. In addition, it poses a growing threat to global public health and requires action by all sectors of government, industry, healthcare professions and society. The success of surgery and medical therapy will increasingly be compromised in the absence of effective antibiotics. On the other hand, the dissemination of multidrug-resistant infections will increase the need for laboratory tests and the use of more expensive drugs, thus increasing the cost of healthcare.

In the science of materials, there is an interest in the development of new antimicrobial therapies, capable of fighting opportunistic infections. For this reason, it is important as a first step, to take into account the different laboratory methods that can be used to evaluate antimicrobial activity *in vitro*. The aim of this methods is to detect possible drug resistance in common pathogens [2], but also which are the most appropriate assays to be used in new agents and materials what that have a potential therapy application.

Nanotechnology and tissue engineering are potential applications for the reported antimicrobial properties of nanoparticles (NPs). Some of the potential advantages of NPs, to fight against microorganisms, are that they do not generate resistance and are a safe potential antimicrobial alternative for clinical use [3–5].

However, the research in this area is needed to understand the mechanism of action of NPs and how to design better therapies. In recent years, nanoparticles have been incorporated into the medical field as an alternative to new antimicrobial agents, especially oxides based in silver (Ag), copper (Cu), and titanium (Ti) [6]. On the other hand, findings have raised concerns about their possible toxic effects in humans, triggering an interest to investigate more about the nanotoxicology and the search for new antibacterial nanomaterials with nontoxic properties for human being [7–9].

The antibacterial activity of the NPs depends of the size and shape; so it requires active research of nanometer-scale materials. Recently, basic and applied research has been done on various metal oxides with different shapes and sizes has carried out for their application in a broad scale of areas such as catalysis, in semiconductors, sensors, controlled release of drugs and as antimicrobial agents.

The physical and chemical properties of metal oxide NPs allow their interaction with biological systems, which has become of vital importance due to the increasing resistance of bacteria. Within these properties, there are shape, size, roughness, zeta potential and coatings, among others [5, 10]. The antimicrobial activity presented by the NPs of metal oxides could have a mainly therapeutic application, but it can also be extended to the food industry, to water purification and to the textile industry.

The present chapter will be focused on recent reports that explore the relationships between the use of NPs in the medical field, with an emphasis on their role as antimicrobial agents and the physicochemical properties of metal oxide NPs that influence their antimicrobial response.

Additionally, findings will address antimicrobial activity of novel metal oxides NPs based in zinc, manganese, iron and magnesium. Also, we will discuss diverse methods for the assessment of antimicrobial activity that can have uses for metal oxide NPs and which complies with quality according to official standards to evaluate antimicrobial agents and materials.

2. Antimicrobial applications of nanoparticles

The rapid emergence of resistant pathogens is occurring worldwide, endangering the efficacy of antibiotics, which have pushed medicine to evolve and save millions of lives. Resistance of the pathogens has been attributed to the overuse and misuse of antibiotic medications, as well as the lack of new strategies for antibacterial development to address the challenge [11, 12]. This challenge suggests that the focus of research on resistance pathogens must be turned to the discovery of novel strategies to fight the pathogen infections. One of the new areas that is emerging in response for this challenging menace is the use of nanotechnology, mainly by the identification of how the manipulation of materials for the synthesis of NPs could be utilized for the therapeutic management of pathogen infections [13].

Research on the synthesis, characterization and application of NPs as an antimicrobial system is a new area of interest in the biomedical and healthcare fields due to the possible enhancement of nanoparticles within their physiochemical behavior against drug-resistant pathogens due to size effect, doping effect, could be cost-effective and they are quite stable enough for long-term storage with a prolonged shelf-life. Moreover, the NPs could be subjected to sterilization by methods of high temperature, gamma irradiation or plasma treatment without losing its properties or inactivation [14].

In reference to the biomedical field, the benefits of nanotechnology have been quite substantial, for example, there are devices with antimicrobial nanoproperties such as heart valves, catheters, and dental implants [10]. The type of nanolayers covering these kind of devices can delay or inhibits the adhesion and growth of bacteria such as *Streptococcus mutans*, *Staphylococcus epidermis*, and *Escherichia coli*. Other implantable material is bone cement based polymethyl methacrylate (PMMA) with Ag nanoparticles that have demonstrated significant reduction in the number of arthroplasty surgery-related infections, including methicillin-resistant *Staphylococcus aureus* (MRSA), *S. aureus*, *S. epidermidis*, and *Acinetobacter baumannii* infections [10, 15].

The application of nanoparticles used to fight against pathogens consist mainly of metals and metal oxides of zinc, silver, copper or titanium, because they naturally exhibit microbiocidal or microbiostatic actions and have demonstrated bactericidal activity against both Gram-positive and Gram-negative bacteria. The bactericidal application of metal NPs is based on the mechanism that affect the respiration system by photocatalytic production of reactive oxygen species (ROS) that damage cellular and viral components that ultimately leads to bacterial death, compromising the bacterial cell wall/membrane, inhibition of enzyme activity and DNA synthesis, interruption of energy transduction and the most important the pathogens do not develop resistance to metal NPs [16, 17].

Besides as metal nanoparticles could target the bacterial cell wall; there is an opportunity to dope the nanoparticles with relevant antibiotics to enhance their antibacterial action through synergy offering multiple advantages as controllable with sustained and relatively uniform distribution release in the target tissue, improving the solubility, minimized side effects and enhanced cellular internalization [14, 18, 19].

3. Metal oxide nanoparticles as antimicrobial agents

The resistance of microorganisms to the action of antimicrobial agents, especially antibiotics, is a serious public health problem, which has been a reason for the search and development of new antimicrobials through nanotechnology.

The manipulation on a nanoscale of metal oxide has provided new research in the pharmaceutical area due to the antimicrobial properties of these oxides, according to data revealed in *in vitro* studies [20, 21]. In this sense, the metal oxide NPs between 1 and 100 nm with different shapes allow their physical and chemical properties could become in some promise antimicrobial agents against infectious diseases for the recent findings about their interaction which has become of vital importance due to the increasing of infection diseases by bacterial resistance [22, 23].

3.1 Mechanisms of antimicrobial activity

There are findings about the potential mechanisms of action, where it attempts to explain the bactericidal effect of metal oxide NPs [10, 24–27]. Some of these include the action of reactive oxygen species (ROS), the electrostatic interaction, accumulation, ions delivered and contact by itself of NPs, that induce a several effects from outside and into the bacteria, and that it will be described below (**Figure 1**).

3.1.1 Formation of reactive oxygen species (ROS)

They are a group of reactive molecules produced in some metabolic processes in which oxygen participates: the superoxide anion O^{2-} which is a powerful oxidizing agent very reactive with water. Hydrogen peroxide H_2O_2 and the hydroxyl radical (•OH) which is the most reactive, since accepting one more electron, gives rise to a water molecule. Metal oxides NPs are capable of producing different reactive oxygen species may participate in different types of reactions in which they can undergo oxidation or reduction processes. ROS produce disruption of DNA, damage by oxidation of polyunsaturated fatty acids and amino acids. The alteration of the balance in the mechanisms of production and elimination of ROS, in favor of production, originates the state of oxidative stress in the bacteria cell. In the case of O_2 and H_2O_2 cause less acute stress reactions and can be neutralized by endogenous antioxidants, such as superoxide and catalase enzymes, while OH^- and O_2 can lead to acute microbial death.

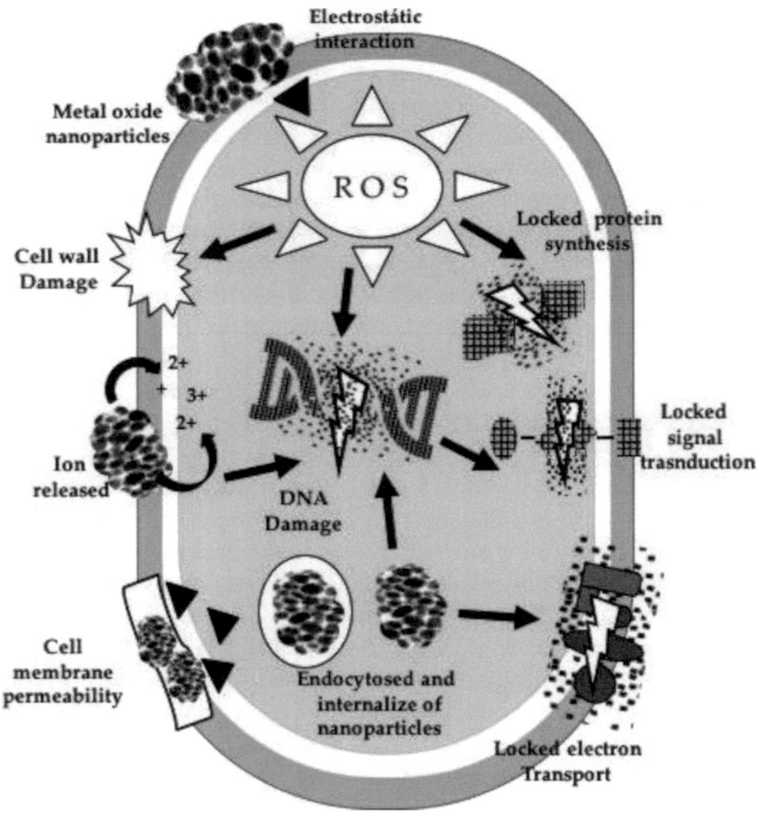

Figure 1.
Mechanisms of action of the bactericidal effect from metal oxide nanoparticles.

3.1.2 Damage to the wall-cell membrane due to electrostatic interaction and accumulation

The electronegative groups of the polysaccharides in the bacterial membrane have an attraction sites by metal cations. The difference in charge between bacterial membranes and the NPs of metal oxides leads to electrostatic attraction and thus accumulates on the bacteria surface, altering the structure and permeability of the cell membrane. Gram-negative bacteria have a higher negative charge than Gram-positive bacteria and therefore the electrostatic interaction will be stronger in Gram-negative strains. The pores of the membranes are in the order of nanometers, therefore the smaller the particle size and the greater the surface area, the greater the efficiency of the metal oxide nanoparticles. In the same way, the cations extracted from the NPs of the metal oxides and their accumulation in the cell wall, create pits in it, leading to a change in permeability due to the sustained release of lipopolysaccharides, membrane proteins and intracellular factors. In addition, this mechanism has been linked to the interruption of the replication of adenosine triphosphate (ATP) and the deoxyribonucleic acid (DNA) of the bacterium, leading to its death. One study indicates that the action of NPs depends on the components and structure of the bacterial cell. The unique components of Gram-negative bacteria, such as LPS, can prevent the adhesion of metal oxides NPs to the barrier of bacterial cells and regulate the flow of ions in and out of the bacterial cell membrane.

3.1.3 Loss of homeostasis by metal ions

The balance of metallic elements is essential for microbial survival, since it regulates metabolic functions by helping coenzymes, cofactors and catalysts. When the bacteria have an excess of metals or metal ions, there will be a disorder in the metabolic functions. Metal ions bind with DNA and alter the helical nature by cross-linking between and within the DNA strands. The metal ions neutralize the charges in LPS and increase the permeabilization of the outer membrane. The ions of metal oxides might also cause the decomposition of bacterial cells due to the diffusion of metal ions by generating large amounts of hydroxyl radicals and diffusion in bacterial cells. Other studies indicate that NPs of metal oxides slowly release metal ions through adsorption, dissolution and hydrolysis; they are toxic and abrasive to bacteria and, therefore, lyse the cells.

3.1.4 Dysfunction of proteins and enzymes

Protein dysfunction is another mode of antibacterial activity exhibited by NPs of metal oxides. The metal ions catalyze the oxidation of the side chains of amino acids resulting in carbonyls bound to proteins. The carboxylation levels within the protein molecule serve as a marker for the oxidative damage of the protein. This carboxylation of proteins will lead to the loss of catalytic activity in the case of enzymes, which finally triggers the degradation of proteins.

3.1.5 Inhibition of the transduction signal

Electrical properties of metal oxide NPs interact with nucleic acids inducing suppress of cell division by altering processes of replication of the chromosomal DNA and the plasmid in microorganism. It is known that signal transduction in bacteria is affected by NPs of metal oxide. Phosphotyrosine is an essential component of mechanism of signal transduction in bacteria. NPs dephosphorylate the phosphotyrosine residues, which inhibits signal transduction and, ultimately, obstructs growth of bacteria.

3.2 Nanoparticle characteristics and their influence of antimicrobial activity

Other factors of the antimicrobial activity, has been sought to analyze what characteristics influence the microbial response to the action of the metal oxide nanoparticles. It is known of existing reports concerning the chemical-physics properties from the metal oxide nanoparticles, but taking into consideration factors like the shape, size, roughness, zeta potential and coatings, etc., that influence the resultant antimicrobial effectiveness [5, 10]. These results could have a mainly therapeutic application in medicine, but it can also be extended to the food industry, to water purification and to the textile industry [28].

3.2.1 Size and shape

Several reports mention that the size and shape are the most important factors to the antimicrobial activity [23, 29–31]. With respect to size there are findings where this is a crucial factor to damage the bacterial systems for many reasons. The sizes as <30 nm are factors that allows the accumulation and penetration into the bacteria causing damage and consequently leading to bacteria death (<10 nm) [23]. The same authors point out that metal oxide nanoparticles with a size greater than 10 nm promotes the permeability when coming into contact with bacteria [23]. In relation with this, the specific surface area by the nanoparticle size affects the surface to mass ratio affecting on surface reactivity. For this reason, they can also have influence in many direct mechanisms of toxicity against the bacteria and the subsequent loss of viability (**Figure 1**).

With respect to shape, it is by knowing that depending on the synthesis method, it will obtain the form of the nanoparticle [32]. Numerous studies shown various forms obtained like spherical, rod-shaped, truncated triangular, nanotubes, nanorods, nanowires, nanosphere, nanoneedles, nanorings and nanocubic [23, 33, 34]. Evidence reports that needle-shaped metal oxides nanoparticles present higher antibacterial activity than cubic shaped, based on the optical and fluorescence intensity [30].

3.2.2 Surface and zeta potential

The relation between the surface nanoparticle/nanomaterial and bacterial adhesion has not been fully studied and there are few reports about it. Some studies report that the adsorption of bacterial proteins is promoting by the surface area-to-mass ratio carry out the reduction in bacterial adhesion [35–37]. Surface nanomaterials have high degree of roughness, therefore bacteria cell membranes cannot adhere to the surface nanomaterial; so the bacteria adhesion is reduced [10, 38, 39].

The surface charge or zeta potential could be another property of the nanoparticle related with bacteria adhesion since it is important to mention that if the surfaces with negative charge are capable to decrease the interaction with bacteria charged negatively, the surface of nanomaterial with negative charge could obtain the same effect, compromising bacterial adhesion [10, 23, 40]. On the other hand, the electrostatic attraction occurs when the nanoparticles are positively charged promoting the accumulation in bacterial cell membrane, which is negatively charged and then they penetrate inside the bacteria triggering other mechanisms [23] (**Figure 1**).

3.2.3 Chemical doping

Nanoparticle chemical doping is a modification and functionalization around the surface of nanoparticles to regulate and control the interaction with bacteria and enhance their antimicrobial effect. Reports have shown this method as a factor to improve the presence of surface oxygen atoms that promote the production of reactive oxygen species (ROS) [23]. Similarly, the chemical functionalization increase of the surface-area-to-volume ratio results in increasing the antimicrobial potential activity [38]. Also, this procedure has prevented the agglomeration and the solubility in different solutions [10].

3.3 Metal oxide nanoparticles

The transition metal oxides (TMO) are compounds with unique electronic properties, most magnetic phenomena involve this type of oxides. The nanostructures formed by TMO, due to their dimensions of a few nanometers and their large surface area, are ideal candidates to interact with bacteria. It is known that NPs of the silver are excellent antimicrobial agents and they are the more studied and reported. However, *in vitro* and *in vivo* studies indicate that nanoparticles based Ag, Cu and Ti are toxic to mammalian cells derived from the skin, liver, lung, brain, vascular system, and also gives rise to a distribution in other organs, where are in accumulations [8, 9].

Therefore, is important that the different metal oxides nanoparticles will be studied and guarantee it clinical use. In particular, there are reports concerning to zinc oxide (ZnO), trimanganese tetroxide (Mn_3O_4), magnetite (Fe_3O_4) and magnesium oxide (MgO) nanoparticles that have antimicrobial properties.

3.3.1 ZnO nanoparticles

Zinc oxide is a compound with excellent antimicrobial properties. It is an n-type semiconductor with a band gap of 3.3 eV. ZnO NPs can adopt a wide variety of morphologies such as; rings, propellers, belts, wires, among others [41, 42]. The antimicrobial activity of ZnO NPs happens by different mechanisms, one of these is the ROS generation [43] inside the cell. It has been proposed that ZnO NPs can act to generate cell death, or the release of Zn^{2+} ions, whose excess generates an alteration of cellular metabolism. Some species reported as susceptible to ZnO nanoparticles are; *S aureus*, *S. epidermidis*, *Streptococcus pyogenes*, *Enterococcus faecalis* [44], *Bacillus subtilis*, *Escherichia coli* and *Klebsiella pneumonia* [41]. These bacteria can generate intra-hospital infections causing serious infectious diseases and some strains are found in water or food, so ZnO NPs can have a possible application in these areas.

3.3.2 Mn_3O_4 nanoparticles

The trimanganese tetroxide, Mn_3O_4, is a mixed oxide of manganese (Mn (II) Mn $(III)_2O_4$) is a normal spinel and crystallizes in cubic form. It occurs in nature as the hausmannite mineral. The antimicrobials properties of Mn_3O_4 NPs have been little studied. Has been reported an effect of these diseases against strains of *Vibrio cholerae*, *Shigella* sp., *Salmonella* sp., and *E. coli* [45]. The effect of the NPs of Mn_3O_4 against has been evaluated against *E. coli* and *S. aureus* through microdilution assays [46]. The results of the minimum inhibitory concentration (MICs) indicated that the bacteria *E. coli* was more sensitive to the action of the NPs of Mn_3O_4. It was observed that the inhibitory effect proportionally increases to the concentration of Mn_3O_4 NPs, which could divide the different characteristics of the surfaces of the bacterial cells and their interaction with the NPs, therefore the mechanism of action could be focused on the bacterial wall membrane.

3.3.3 Fe_2O_3 nanoparticles

Iron oxide (III) is a very stable oxide, it crystallizes in hexagonal form and is found in nature as the mineral hematite α-Fe_2O_3. The nanostructures of this oxide take different forms as they are nanowires, nanotubes, nanospheres, etc. [47]. Although its synthesis has been widely studied, its possible antibacterial effect not. The Fe_2O_3 NPs bactericidal effect against *E. coli* and *S. aureus* has been reported, where an increase of this effect is observed, as the concentration of iron oxide NPs increases [48]. A bactericidal effect has also been seen on *P. aeruginosa* with a minimum inhibitory concentration of 0.06 mg/L [49]. Another study reports on the bactericidal activity of nanostructured hematite against a variety of Gram-positive and Gram-negative bacteria; *P. aeruginosa*, *S. aureus*, *K. pneumoniae*, *Lysinibacillus sphaericus* and *Bacillus safensis* [50]; proposing some mechanisms of action depending on the activity observed in each stage of the growth of the bacteria in question. A bactericidal effect of NPs of Fe_2O_3 against *S. epidermidis* has even been determined [51].

From its properties, its possible application in the remediation of the environment and water, as well as in the biomedical area, has been proposed, due to the different studies of cytotoxicity that have been carried out [47].

3.3.4 MgO nanoparticles

Magnesium oxide is in nature as the mineral periclase [52]. The antibacterial activity of MgO against Gram-positive and Gram-negative bacteria has been reported. It has been proposed that MgO NPs can damage the cell membrane causing the loss of intracellular contents and causing the death of bacterial cells [53]. The generation of reactive oxygen species has been attributed to the surface alkalinity of the MgO NPs [54]. The antibacterial activity of NPs of MgO against Gram-negative bacteria has been evaluated; *E. coli* and *P. aeruginosa* (500 and 1000 μg/mL) and in a Gram-positive bacterium; *S. aureus* (1000 μg/mL) [55]. The MgO NPs potentiated lipid peroxidation induced by ultrasound in the liposomal membrane. In this case the mechanism of action could be associated to the presence of defects, or to the lack of oxygen on the surface of the NP, leading to lipid peroxidation and the generation of reactive oxygen species [55]. The antibacterial effect and mechanism of action of NPs of MgO against strains of *Campylobacter jejuni*, *E. coli* and *Salmonella enteritidis* has been studied [56]. In this case, it was observed that the permeability of the bacteria's membrane, after exposure to the MgO NPs, was compromised, finding the presence of hydrogen peroxide that would subsequently cause cell death. Studies of *P. aeruginosa* and *S. aureus* versus MgO NPs showed a greater zone of inhibition in *S. aureus* than in *P. aeruginosa* [56]. Based on previous work, the authors note that the bactericidal action of MgO NPs may be due to the binding of surface oxygen to bacteria. As the surface area of the particles increases, the concentration of oxygen ions on the surface increases, which results in a more effective destruction of the cytoplasmic membrane and the cell wall of the bacteria.

4. *In vitro* methods for antimicrobial evaluation of nanoparticles based metal oxide

Bacteria exposed to antimicrobials are under selective pressure to evolve and adapt, this natural process leads to antimicrobial resistance. Human kind is facing the growing threat of rapid evolution and dissemination of bacteria resistant to multiple antibiotics. There is, therefore, an urgent need to develop new antimicrobials [57, 58].

Antimicrobial agents include disinfectants, antiseptics, and antibiotics. New agents must be exhaustively tested for efficacy and safety. Evidence-based selection of the microorganisms and the evaluation system is of paramount importance for adequate interpretation of the test results, and for extrapolating from *in vitro* to real-life scenarios.

The use of nanoparticles especially based in metal oxides emerge as new antimicrobial agents, therefore it is necessary to test the efficacy of nano-antimicrobials against representative bacterial species. One known limitation of the testing systems currently in use, is that formulations are often challenged *in vitro* with one microbial species at the time, and rarely against multi-species biofilms.

4.1 Regulatory testing

Regulatory agencies require adherence to well established evaluation systems. Regulatory tests applicable to disinfectants, antiseptics or therapeutic antimicrobials vary greatly; and could include to nanoparticles with potential use as antimicrobials.

4.1.1 Disinfectants

When evaluating chemical disinfectants, bacterial endospores are considered the microbial life-form hardest to kill, followed in descendent order by mycobacteria, bacteria in vegetative form, and viruses.

In the United States, the Food and Drug Administration (FDA) regulates chemical sterilants and high level disinfectants (HLD) that are used to reprocess medical instruments [59]. The AOAC International sporicidal and tuberculocidal tests are the accepted methods for evaluation. For a liquid chemical sterilant, the FDA standard tolerates no failures in the AOAC sporicidal test 966.04, and accepts no survivors in simulated-use testing with a challenge inoculum of six logs of spores. The FDA defines HLD as sterilants used under the same contact conditions but for only the contact time needed to reduce *Mycobacterium bovis* in 6 $\log_{10}$ in the tuberculocidal test 965.12 [60]. Moreover, to be approved, the disinfectants should be subjected to worse case scenarios, such as the presence of organic or inorganic contamination, and under simulated use conditions.

In Europe, the CEN/TC 216 technical committee produces current and future disinfectant testing standards [61]. Standard EN-14885-2006 indicates test methods to be used to substantiate claims for products intended for instrument disinfection [62], including mycobacterial/tuberculocidal (EN-14348, EN 14563), bactericidal (EN-13727, EN-14561) and fungicidal (EN-13624, EN-14562) activity tests but the terms "sterility, sterile, sterilization, sterilant" fall outside the scope of CEN/TC 216.

In the US, high, intermediate and low level disinfectants, are regulated by the Environment Protection Agency (except HLD intended to reprocess medical instruments, which fall under FDA's jurisdiction). Intermediate level disinfectants must be tuberculocidal. Products effective only against vegetative bacteria and viruses are regarded as low level disinfectants [63].

Special testing procedures may be applicable to some pathogens of epidemiological interest, such as *Clostridium difficile* [64]. Disinfectants are intended for use on inanimate surfaces. In general, their high concentration precludes their use on living tissues.

4.1.2 Antiseptics

Antiseptics are antimicrobials intended for use on skin and mucous membranes. The same as with low level disinfectants, antiseptics are tested against *Escherichia coli*, *Staphylococcus aureus* and *Pseudomonas aeruginosa* in vegetative form. In the US, antiseptics are regulated by the FDA.

4.2 Antimicrobial activity tests

A relevant test microorganism is chosen: preferably a strain from the American Type Culture Collection (ATCC) or a similar repository. Although, wild-type bacteria from clinical samples also have been used. All necessary controls must be included to assess test reliability and reproducibility. Also, it is important to differentiate between kill and inhibition of growth.

The antimicrobial capability of nanoparticles has been explored by this techniques due studies have suggested that NPs are excellent microbicidal activity [16, 65]. The *in vitro* tests described below are the ones that the most have been used and the regulatory agencies recommend to determine antimicrobial activity of chemical formulations and can be used in the studies of nano-antimicrobials. The use of such tests depends on the objectives and the type of information it want to obtain.

A first approach if nanoparticle has antimicrobial activity is to conduct an antimicrobial activity test, such as a disc diffusion test.

4.2.1 Disk-diffusion method

Mueller-Hinton agar (pH 7.2–7.4) is the culture medium of choice. To standardize disc diffusion, the agar is poured into either Petri dish to only 4 mm in depth, as indicated in the Clinical and Laboratory Standards Institute method [66].

The bacteria are suspended to a 0.5 McFarland turbidity standard equivalent to 150 × 106 cfu/mL. From this suspension, 100 μL are uniformly spread onto the agar. Filter-paper discs 6 mm diameter, containing the test nano-antimicrobial, will be placed over the seeded agar (alternatively, a 50–100 μL well, punched into the agar, will contain the test antimicrobial). After overnight incubation at 37°C, the plates will be examined to assess inhibition rings around the disc.

The size of the nanoparticle, its rate of diffusion, the agar's porosity, and possible charge interactions between the antimicrobial and the agar may affect diffusion and the final size of the inhibition zone. In theory, the highest concentrations will be near the antimicrobial-containing disc and will be diluted away from the center (**Figure 2**).

4.2.2 Agar dilution method

This method is the gold standard for assessing the minimal inhibitory concentration (MIC) [67]. In this method, the melted agar is mixed to contain serial dilutions of the nano-antimicrobial. The resulting antimicrobial containing medium is plated into Petri dishes. An aliquot containing 10^4 cfu of the test microorganism is placed onto the agar's surface, and incubated overnight. Then, the plates will be examined for growth to determine the last effective concentration to inhibit growth (**Figure 3**).

4.2.3 Broth dilution method

This method is often used because is more versatile and less laborious than the agar dilution method. Its microtiter plate version (broth microdilution), allows for testing more microorganisms against diverse concentrations of nano-antimicrobials, and can be automated.

Test tubes or wells in a microtiter plate, are prepared with bacteriological broth containing serial dilutions of the test nano-antimicrobial, and seeded with

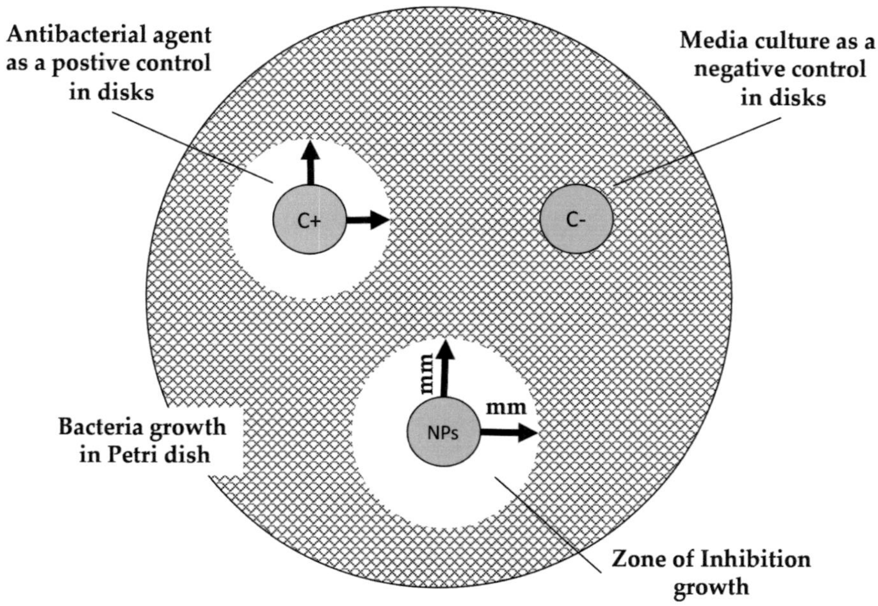

Figure 2.
Disk-diffusion method with NPs.

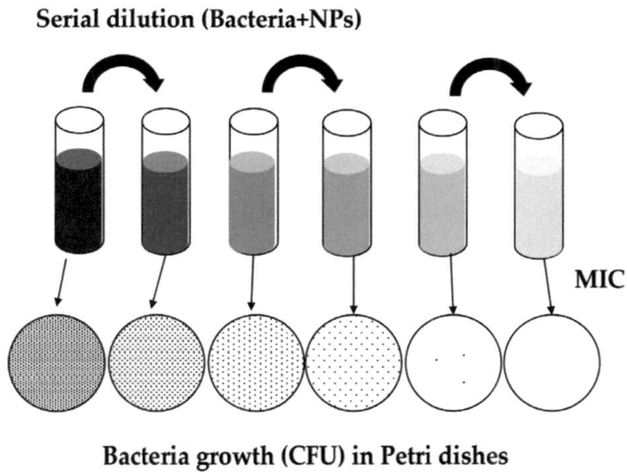

Figure 3.
Agar dilution method with NPs.

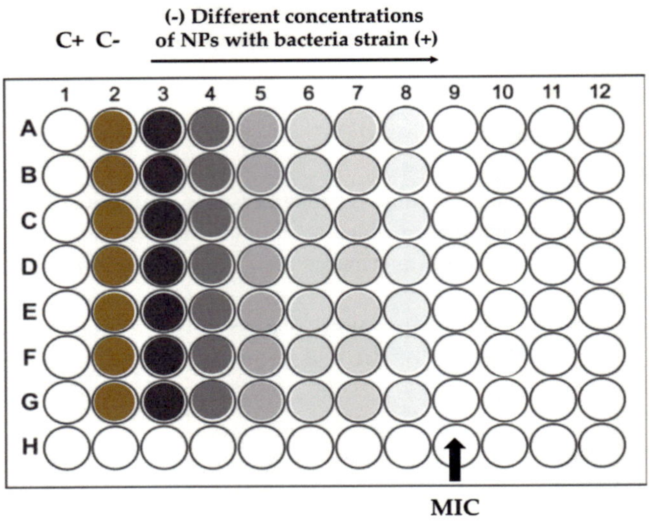

Figure 4.
Broth dilution method with NPs.

bacteria. After, overnight incubation, the tubes or wells are inspected for growth. The lowest concentration of nanoparticles that results in no-growth is the MIC [68] (**Figure 4**).

4.2.4 Time-kill method

After adding the test formulation to a broth culture, antimicrobial activity can be assessed *in vitro* by collecting sequential samples to count survivors. Time-kill allows the assessment of *in-vitro* synergy or antagonism between nano-antimicrobials.

For the time-kill experiments, Mueller Hinton broth is prepared with serial dilutions of the test antimicrobial, alone or in combination. The nano-antimicrobial concentrations may span a range above and below the formulation's MIC, previously obtained from agar dilution tests. Broths are then inoculated with 106 cfu/mL and incubated overnight at 37°C. From time 0 when bacteria are first exposed to the test antimicrobial, samples are obtained at 30 min intervals for up to 6 h. The samples are then plated on nutrient agar. After incubation overnight at 37°C, survivor counts are plotted to obtain a 'time-kill curve' [69].

5. Conclusions

Taking into account the future development and applications of the metal oxides nanoparticles in medicine, a constant search as emergent antimicrobial agents is required, due to the increase of diseases caused by microorganisms resistant to the action of antimicrobial agents as antibiotics.

Implementation of the metal oxides nanoparticles as an alternative to combat bacterial resistance due to increased findings in the mechanisms by which they act, have been the key to a better understanding and approach about the effect and kinetics that metal oxides nanoparticles have on microbial strains.

For this reason it is necessary to establish guidelines and quality standards to research nano-antimicrobials given the fact that there are many alternative methods *in vitro* testing to achieve this objective, but some of them present limitations, while searching for new methods could be able to present specific results that allow us to compare them with *in vivo* testing.

Acknowledgements

This work was financially supported by Facultad de Odontología, División de Estudios de Posgrado e Investigación and Instituto de Ciencias Aplicadas y Tecnología in Universidad Nacional Autónoma de México.

Conflict of interest

The authors declare no conflicts of interest.

Author details

Alejandro L. Vega-Jiménez[1*], América R. Vázquez-Olmos[2], Enrique Acosta-Gío[3] and Marco Antonio Álvarez-Pérez[1]

1 División de Estudios de Posgrado e Investigación, Facultad de Odontología, Laboratorio de Bioingeniería de Tejidos, Universidad Nacional Autónoma de México, Mexico

2 Grupo de Sistemas híbridos y Nanoespectroscopía, Instituto de Ciencias Aplicadas y Tecnología, Universidad Nacional Autónoma de México, Mexico

3 División de Estudios de Posgrado e Investigación, Facultad de Odontología, Laboratorio de Microbiología y Control de Infecciones, Universidad Nacional Autónoma de México, Mexico

*Address all correspondence to: dr.vegalex@gmail.com

References

[1] World Health Organization. Antimicrobial resistance. 2018. Available from: http://www.who.int/mediacentre/factsheets/fs194/en/ [Accessed: October 15, 2017]

[2] Reller LB et al. Antimicrobial susceptibility testing: A review of general principles and contemporary practices. Clinical Infectious Diseases. 2009;**49**(11):1749-1755. DOI: 10.1086/647952

[3] Jan T et al. Synthesis, physical properties and antibacterial activity of metal oxides nanostructures. Materials Science in Semiconductor Processing. 2014;**21**:154-160. DOI: 10.1016/j.mssp.2014.01.006

[4] Kalyani RL, Venkatraju J, Kollu P, Rao NH, Pammi SVN. Low temperature synthesis of various transition metal oxides and their antibacterial activity against multidrug resistance bacterial pathogens. Korean Journal of Chemical Engineering. 2015;**32**(5):911-916. DOI: 10.1007/s11814-014-0262-5

[5] Mamonova I et al. Biological activity of metal nanoparticles and their oxides and their effect on bacterial cells. Nanotechnologies in Russia. 2015;**10**(1-2):128-134. DOI: 10.1134/S1995078015010139

[6] Ling D, Hyeon T. Chemical design of biocompatible iron oxide nanoparticles for medical applications. Small. 2013;**9**(9-10):1450-1466. DOI: 10.1002/smll.201202111

[7] Karlsson HL et al. Copper oxide nanoparticles are highly toxic: A comparison between metal oxide nanoparticles and carbon nanotubes. Chemical Research in Toxicology. 2008;**21**(9):1726-1732. DOI: 10.1021/tx800064j

[8] Jeng HA, Swanson J. Toxicity of metal oxide nanoparticles in mammalian cells. Journal of Environmental Science and Health Part A. 2006;**41**(12):2699-2711. DOI: 10.1080/10934520600966177

[9] Adamcakova-Dodd A, Thorne PS. Grassian VH: In Vivo Toxicity Studies of Metal and Metal Oxide Nanoparticles. In Handbook of Systems Toxicology. Edited by: Daniel A Cascinao, Saura C Sahu. Chichester, UK: John Wiley & Sons, Ltd; 2011:803-834. DOI: 10.1002/9780470744307.gat244

[10] Wang L, Hu C, Shao L. The antimicrobial activity of nanoparticles: Present situation and prospects for the future. International Journal of Nanomedicine. 2017;**12**:1227. DOI: 10.2147/IJN.S121956

[11] Livermore DM. The 2018 Garrod lecture: Preparing for the black swans of resistance. The Journal of Antimicrobial Chemotherapy. 2018;**73**(11):2907-2915. DOI: 10.1093/jac/dky265

[12] Ventola CL. The antibiotic resistance crisis: Part 1: Causes and threats. Pharmacy and Therapeutics. 2015;**40**(4):277-283. PMID: 25859123

[13] Aslam B et al. Antibiotic resistance: A rundown of a global crisis. Infection and Drug Resistance. 2018;**11**:1645-1658. DOI: 10.2147/IDR.S173867

[14] Huh AJ, Kwon YJ. Nanoantibiotics: A new paradigm for treating infectious diseases using nanomaterials in the antibiotics resistant era. Journal of Controlled Release. 2011;**156**(2):128-145. DOI: 10.1016/j.jconrel.2011.07.002

[15] Prokopovich P et al. Potent antimicrobial activity of bone cement encapsulating silver nanoparticles capped with oleic acid. Journal of Biomedical Materials Research. Part B, Applied Biomaterials. 2015;**103**(2):273-281. DOI: 10.1002/jbm.b.33196

[16] Seil JT, Webster TJ. Antimicrobial applications of nanotechnology: Methods and literature. International Journal of Nanomedicine. 2012;**7**: 2767-2781. DOI: 10.2147/IJN.S24805

[17] Díez-Pascual AM. Antibacterial activity of nanomaterials. (Basel). 2018;**8**(6):1-6. DOI: 10.3390/nano8060359

[18] Gao W et al. Nanoparticle-based local antimicrobial drug delivery. Advanced Drug Delivery Reviews. 2018;**127**:46-57. DOI: 10.1016/j.addr.2017.09.015

[19] Qidwai A et al. Advances in biogenic nanoparticles and the mechanisms of antimicrobial effects. Indian Journal of Pharmaceutical Sciences. 2018;**80**(4):592-603. DOI: 10.4172/pharmaceutical-sciences.1000398

[20] Dizaj SM et al. Antimicrobial activity of the metals and metal oxide nanoparticles. Materials Science and Engineering: C. 2014;**44**:278-284. DOI: 10.1016/j.msec.2014.08.031

[21] Hoseinzadeh E et al. Sensitivity coefficient and death kinetics of *Escherichia coli* and *Staphylococcus aureus* to zinc oxide and copper oxide nanoparticles. Journal of Isfahan Medical School. 2012;**30**:200

[22] Hajipour MJ et al. Antibacterial properties of nanoparticles. Trends in Biotechnology. 2012;**30**(10):499-511. DOI: 10.1016/j.tibtech.2012.06.004

[23] Hoseinzadeh E et al. A review on nano-antimicrobials: Metal nanoparticles, methods and mechanisms. Current Drug Metabolism. 2017;**18**(2):120-128. DOI: 10.2174/138920021766616120111146

[24] Hemeg HA. Nanomaterials for alternative antibacterial therapy. International Journal of Nanomedicine. 2017;**12**:8211. DOI: 10.2147/IJN.S132163

[25] Lemire JA, Harrison JJ, Turner RJ. Antimicrobial activity of metals: Mechanisms, molecular targets and applications. Nature Reviews Microbiology. 2013;**11**(6):371. DOI: 10.1038/nrmicro3028

[26] Raghunath A, Perumal E. Metal oxide nanoparticles as antimicrobial agents: A promise for the future. International Journal of Antimicrobial Agents. 2017;**49**(2):137-152. DOI: 10.1016/j.ijantimicag.2016.11.011

[27] Stankic S et al. Pure and multi metal oxide nanoparticles: Synthesis, antibacterial and cytotoxic properties. Journal of Nanobiotechnology. 2016;**14**(1):73. DOI: 10.1186/s12951-016-0225-6

[28] Vázquez Olmos AR, Vega Jiménez AL, Paz Díaz B. Mecanosíntesis y efecto antimicrobiano de óxidos metálicos nanoestructurados. Mundo Nano. Revista Interdisciplinaria en Nanociencia y Nanotecnología. 2018;**11**(21):29-44. DOI: 10.22201/ceiich.24485691e.2018.21.62545

[29] Azam A et al. Antimicrobial activity of metal oxide nanoparticles against Gram-positive and Gram-negative bacteria: A comparative study. International Journal of Nanomedicine. 2012;**7**:6003. DOI: 10.2147/IJN.S35347

[30] Kim DH et al. Effect of the size and shape of silver nanoparticles on bacterial growth and metabolism by monitoring optical density and fluorescence intensity. Biotechnology and Bioprocess Engineering. 2017;**22**(2):210-217. DOI: 10.1007/s12257-016-0641-3

[31] Gao M et al. Controlled synthesis of Ag nanoparticles with different morphologies and their antibacterial

properties. Materials Science and Engineering: C. 2013;**33**(1):397-404. DOI: 10.1016/j.msec.2012.09.005

[32] Somorjai GA, Park JY. Colloid science of metal nanoparticle catalysts in 2D and 3D structures. Challenges of nucleation, growth, composition, particle shape, size control and their influence on activity and selectivity. Topics in Catalysis. 2008;**49**(3-4):126-135. DOI: 10.1007/s11244-008-9077-0

[33] Simon-Deckers A et al. Size-, composition-and shape-dependent toxicological impact of metal oxide nanoparticles and carbon nanotubes toward bacteria. Environmental Science & Technology. 2009;**43**(21):8423-8429. DOI: 10.1021/es9016975

[34] Stoimenov PK et al. Metal oxide nanoparticles as bactericidal agents. Langmuir. 2002;**18**(17):6679-6686. DOI: 10.1021/la0202374

[35] Sukhorukova I et al. Toward bioactive yet antibacterial surfaces. Colloids and Surfaces B: Biointerfaces. 2015;**135**:158-165. DOI: 10.1016/j.colsurfb.2015.06.059

[36] Song Y, Chen L. Effect of net surface charge on physical properties of the cellulose nanoparticles and their efficacy for oral protein delivery. Carbohydrate Polymers. 2015;**121**:10-17. DOI: 10.1016/j.carbpol.2014.12.019

[37] Young J-j et al. Positively and negatively surface-charged chondroitin sulfate-trimethylchitosan nanoparticles as protein carriers. Carbohydrate Polymers. 2016;**137**:532-540. DOI: 10.1016/j.carbpol.2015.10.095

[38] Ben-Sasson M et al. Surface functionalization of thin-film composite membranes with copper nanoparticles for antimicrobial surface properties. Environmental Science and Technology. 2013;**48**(1):384-393. DOI: 10.1021/es404232s

[39] Anselme K et al. The interaction of cells and bacteria with surfaces structured at the nanometre scale. Acta Biomaterialia. 2010;**6**(10):3824-3846. DOI: 10.1016/j.actbio.2010.04.001

[40] Pan X et al. Investigation of antibacterial activity and related mechanism of a series of nano-$Mg(OH)_2$. ACS Applied Materials and Interfaces. 2013;**5**(3):1137-1142. DOI: 10.1021/am302910q

[41] Król A et al. Zinc oxide nanoparticles: Synthesis, antiseptic activity and toxicity mechanism. Advances in colloid and interface science. 2017;(249):37-52. DOI: 10.1016/j. cis.2017.07.033

[42] Wang ZL. Nanostructures of zinc oxide. Materials Today. 2004;**7**(6):26-33. DOI: 10.1016/S1369-7021(04)00286-X

[43] Kaftelen H et al. EPR and photoluminescence spectroscopy studies on the defect structure of ZnO nanocrystals. Physical Review B. 2012;**86**(1):014113-9. DOI: 10.1103/PhysRevB.86.014113

[44] Santhoshkumar J, Kumar SV, Rajeshkumar S. Synthesis of zinc oxide nanoparticles using plant leaf extract against urinary tract infection pathogen. Resource-Efficient Technologies. 2017;**3**(4):459-465. DOI: 10.1016/j.reffit.2017.05.001

[45] Chowdhury A-N et al. Oxidative and antibacterial activity of Mn_3O_4. Journal of Hazardous Materials. 2009;**172**(2):1229-1235. DOI: 10.1016/j.jhazmat.2009.07.129

[46] Azhir E et al. Preparation, characterization and antibacterial activity of manganese oxide

nanoparticles. Physical Chemistry Research. 2015;**3**(3):197-204. DOI: 10.22036/pcr.2015.9329

[47] Tadic M et al. Synthesis of core-shell hematite (α-Fe_2O_3) nanoplates: Quantitative analysis of the particle structure and shape, high coercivity and low cytotoxicity. Applied Surface Science. 2017;**403**:628-634. DOI: 10.1016/j.apsusc.2017.01.115

[48] Rufus A, Sreeju N, Philip D. Synthesis of biogenic hematite (α-Fe_2O_3) nanoparticles for antibacterial and nanofluid applications. RSC Advances. 2016;**6**(96):94206-94217. DOI: 10.1039/C6RA20240C

[49] Irshad R et al. Antibacterial activity of biochemically capped iron oxide nanoparticles: A view towards green chemistry. Journal of Photochemistry and Photobiology B: Biology. 2017;**170**:241-246. DOI: 10.1016/j.jphotobiol.2017.04.020

[50] Muthukumar H et al. Iron oxide nano-material: Physicochemical traits and in vitro antibacterial propensity against multidrug resistant bacteria. Journal of Industrial and Engineering Chemistry. 2017;**45**:121-130. DOI: 10.1016/j.jiec.2016.09.014

[51] Groiss S et al. Structural characterization, antibacterial and catalytic effect of iron oxide nanoparticles synthesised using the leaf extract of *Cynometra ramiflora*. Journal of Molecular Structure. 2017;**1128**:572-578. DOI: 10.1016/j.molstruc.2016.09.031

[52] Rankin DW. CRC Handbook of Chemistry and Physics. In: David R, editor. Lide. Boca Raton: CRC (Taylor and Francis Group); 2009. DOI: 10.1021/ja069813z

[53] Jin T, He Y. Antibacterial activities of magnesium oxide (MgO) nanoparticles against foodborne pathogens. Journal of Nanoparticle Research. 2011;**13**(12):6877-6885. DOI: 10.1007/s11051-011-0595-5

[54] Yamamoto O et al. Antibacterial characteristics of $CaCO_3$–MgO composites. Materials Science and Engineering: B. 2010;**173**(1):208-212. DOI: 10.1016/j.mseb.2009.12.007

[55] Krishnamoorthy K et al. Antibacterial activity of MgO nanoparticles based on lipid peroxidation by oxygen vacancy. Journal of Nanoparticle Research. 2012;**14**(9):1063. DOI: 10.1007/s11051-012-1063-6

[56] Bindhu M et al. Structural, morphological and optical properties of MgO nanoparticles for antibacterial applications. Materials Letters. 2016;**166**:19-22. DOI: 10.1016/j.matlet.2015.12.020

[57] Brooks BD, Brooks AE. Therapeutic strategies to combat antibiotic resistance. Advanced Drug Delivery Reviews. 2014;**78**:14-27. DOI: 10.1016/j.addr.2014.10.027

[58] Gottlieb S. FDA's Strategic Approach for Combating Antimicrobial Resistance. September 14, 2018. Washington, DC. Available from: https://www.fda.gov/NewsEvents/Speeches/ucm620495.htm

[59] US Food and D. Administration, Guidance for Industry and FDA Reviewers: Content and Format of Premarket Notification [510 (k)] Submissions for Liquid Chemical Sterilants/High Level Disinfectants. Rockville, MD: US Department of Health and Human Services, Food and Drug Administration; 2000. pp. 8-9. Available from: www.fda.gov/medicaldevices/deviceregulationandguidance/

guidancedocuments/ucm073773.htm [Accessed: November 17, 2018]

[60] AOAC International. Official Methods of Analysis of the Official Analytical Chemists. Arlington, VA: Ed: Kenneth Helrich, AOAC Inc.; 1990. pp. 133-146 (chapter 6)

[61] Holah JT. CEN/TC 216: Its role in producing current and future European disinfectant testing standards. International Biodeterioration and Biodegradation. 2003;**51**:239-243. Available from: https://standards.cen.eu/dyn/www/f?p=204:7:0::::FSP_ORG_ID:6197&cs=1C21F982635037747B259BDC2783DB513

[62] Comité Européen de Normalisation. EN 14885:2006 Chemical disinfectants and antiseptics—Application of European Standards for chemical disinfectants and antiseptics. Brussels, Belgium. 2015

[63] US Environmental Protection Agency. Product Performance Test Guidelines OCSPP 810.2000: General Considerations for Testing Public Health Antimicrobial Pesticides. Guidance for Efficacy Testing. EPA 712-C-17-002. Washington, D.C. February; 2018

[64] US Environmental Protection Agency. EPA MLB SOP-MB-31: Procedure for the OECD Quantitative Method for Testing Antimicrobial Products against Spores of Clostridium difficile (ATCC 43598) on Inanimate, Hard, Non-porous Surfaces. Washington, D.C. December; 2017

[65] Hoseinzadeh E, Makhdoumi P, Taha P, Hossini H, Pirsaheb M, Omid Rastegar S, et al. A review of available techniques for determination of nano-antimicrobials activity. Toxin Reviews. 2017;**36**(**1**):18-32. DOI: 10.1080/15569543.2016.1237527

[66] CLSI. Performance Standards for Antimicrobial Disk Susceptibility Tests; Approved Standard-Eleventh Edition. Clinical and Laboratory Standards Institute document M02-A11. Wayne, PA; 2012

[67] Baker CN et al. Comparison of the E test to agar dilution, broth microdilution, and agar diffusion susceptibility testing techniques by using a special challenge set of bacteria. Journal of Clinical Microbiology. 1991;**29**(3):533-538. PMC269813/

[68] Wiegand I, Hilpert K, Hancock RE. Agar and broth dilution methods to determine the minimal inhibitory concentration (MIC) of antimicrobial substances. Nature Protocols. 2008;**3**(2):163-175. DOI: 10.1038/nprot.2007.521

[69] MacGowan A et al. A new time-kill method of assessing the relative efficacy of antimicrobial agents alone and in combination developed using a representative β-lactam, aminoglycoside and fluoroquinolone. Journal of Antimicrobial Chemotherapy. 1996;**38**(2):193-203. DOI: 10.1093/jac/38.2.193

Chapter 3

Nanoformulated Delivery Systems of Essential Nutraceuticals and Their Applications

Lebogang Katata-Seru, Bathabile Ramalapa and Lesego Tshweu

Abstract

Malnutrition and poor diet constitute the number one driver of the global burden of disease. Undernutrition is responsible for up to 50% of all deaths in children under the age of 5. In South Africa, 25% of the country's children suffer from undernutrition. This increases the risk of child mortality as well as contracting infectious diseases. It also affects the physical and intellectual development of the children. The greatest drawback in malnutrition is the deficiency of essential nutraceuticals involved in important biological functions. Innovative technologies such as nanoformulated products are needed for food and agriculture in order to enhance the children's health. The evaluation and application of various nanoformulated delivery systems will be explored for improving the stability and bioavailability of essential nutraceuticals for consumers.

Keywords: nanoformulated, nutraceuticals, delivery systems, undernutrition, food, agriculture

1. Introduction

Over the past years, nutraceuticals have been explored as novel medicinal dietary products in the food and pharmaceutical industries. Dr. Stephen De Felice invented "nutraceutical" as a term in an attempt to promote medical health research [1–2]. He defined it as a food or part of a food that provides medical or health benefits, including the prevention and treatment of disease. Nutraceuticals are mostly classified into three broad groups such as dietary supplements (glucosamine, probiotics, etc.), herbals (herbs or botanical products), and nutrients (vitamins, minerals, etc.). In addition, they are consumed daily by human beings as an alternative to modern medicine, thus promoting quality life and increasing life expectancy. They have proved to offer benefits such as acting as a natural antioxidant and immune booster, fewer side effects than drugs, improved bioavailability, and long half-life [3]. The focus of the chapter will be limited to various sources of vitamins especially nanoformulated liposoluble vitamins.

Great progress toward enhancing the stability and bioavailability of vitamins, thus promoting health benefits among consumers, has been achieved by various researchers [4–6]. **Figure 1** shows a classification of vitamins according to their solubility. In the last decade, an extensive amount of research has focused on the

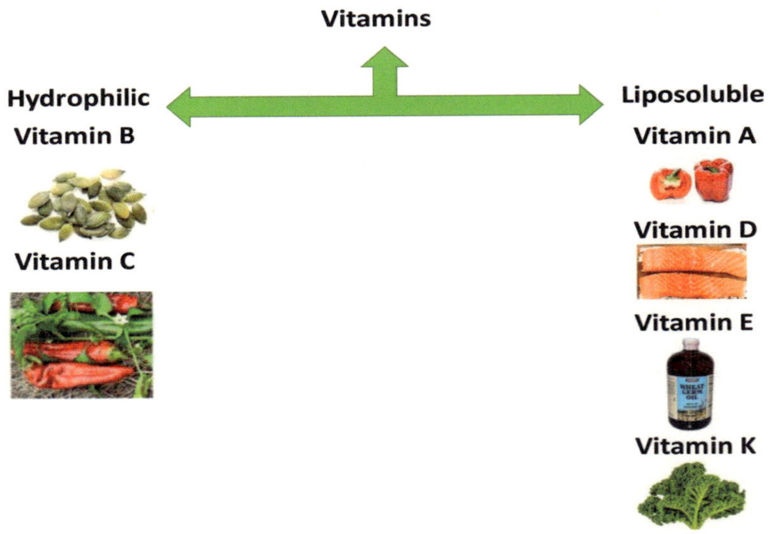

Figure 1.
Classification of vitamins with some examples of high rich foods.

hydrophilic vitamins as compared to liposoluble ones [7, 8]. Different approaches have been explored to improve the stability and functionality of hydrophilic vitamins during product development and storage of food because of the exposure to high temperature, oxygen, and light. Novel methods were used by Alishahi et al. [9] to increase the shelf life and delivery of vitamin C using chitosan nanoparticles. In addition, three different vitamins (B9, B12, and C) were successfully encapsulated in water-soluble derivatives of chitosan biopolymer [10]. Their study showed that N,N,N-trimethyl chitosan nanoparticles can successfully be used as a stable vitamin carrier system with potential applications in foodstuffs.

Table 1 highlights some of their health benefits and disease prevention [11–13]. Many of these vitamins have been found to have major limitations such as low chemical stability, sensitive to oxidation, high melting points, and poor solubility, thus leading to low bioavailability [14]. The inability of the human body to produce vitamins forces humans to have a balanced diet in order to intake the recommended supply of essential nutrients. The inadequate intake of various vitamins through diet may compromise biological functions such as vision, growth and development, immunological activity, reproduction, and cellular growth [11]. The application of adequate treatment can reduce the risk of development of complications related to deficiency of vitamins. Food fortification and supplements have been used as strategies to prevent vitamin deficiency. In order for nutritional supplement premixes and fortified food to work, the vitamins and micronutrients contained in these products need to remain active until consumption, which may not always be the case. Premixes and fortified foods may lose a large percentage of vitamins and micronutrient activity before consumption via processing, packaging, transportation, and storage. Therefore there is an urgent need to develop and explore cost-effective innovative approaches that will improve the stability of nutraceuticals especially liposoluble vitamins.

The delivery of vitamins using nanotechnology has attracted a number of attention recently [14] and is proposed as one of the possible innovative approaches. Numerous methods including spray-drying, spray-cooling, phase separation, emulsion systems, liposome solid lipid nanoparticles (SLN), and inclusion complexation have been proposed for the nanoformulations of liposoluble vitamins as depicted in **Figure 2**. Some of the methods including nanoemulsions, polymeric and lipid nanoparticles, etc. will be discussed in the chapter.

Vitamin, chemical name	Recommended daily intake	Health benefits	Deficiency
Vitamin A, retinol	300 µg (1–3 years), 400 µg (>4 years)	Antioxidant, improves vision and bone growth, regulates cell proliferation, treatment of skin cancer and leukemia	Xerophthalmia caused by unbalanced diet or malabsorption diseases, liver disorders, night blindness
Vitamin D, cholecalciferol	5 µg (4–8 years)	Essential for healthy bones, controls the amount of calcium in the blood	Growth slowed in children, inadequate diet, inadequate exposure to sunlight, severe deficiency leads to rickets, osteomalacia, muscle pain, severe asthma in kids
Vitamin E, tocopherol	6 mg (1–3 years), 7 mg (4–8 years)	Strengthens body's immune system, keeps blood vessels clear and flowing well, boosts the immune system, healthy skin and eyes, antioxidant	Deficiency might lead to premature infants
Vitamin K, phylloquinone	30 µg (1–3 years), 55 µg (4–8 years)	Essential for blood clotting and blood coagulation	Subdermal hemorrhaging, deficiency leads to uncontrolled bleeding and clotting

Table 1.
Recommended daily intake, Health benefits, and deficiency of liposoluble vitamins [11–13].

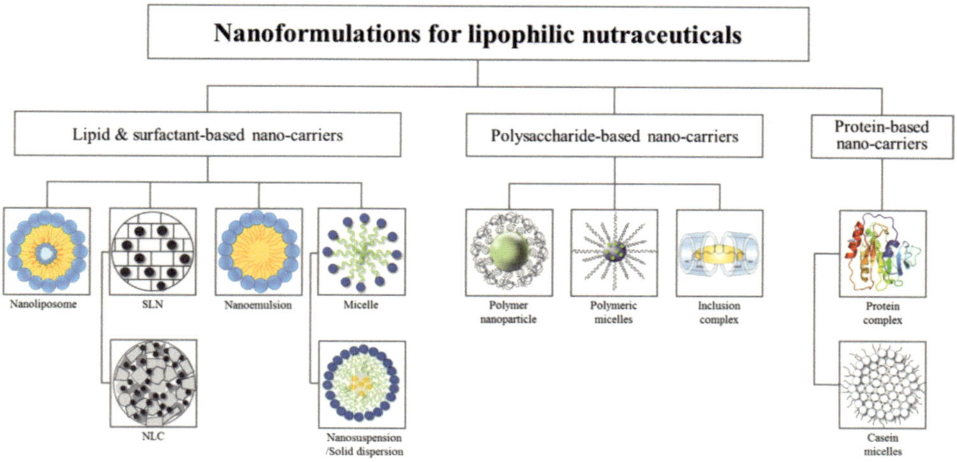

Figure 2.
Nanoformulations for liposoluble vitamins [14].

1.1 Nanoemulsions

Colloidal dispersions consisting of oil droplets dispersed in an aqueous medium in the 5–200 nm range are known as a nanoemulsion. They are isotropic systems, which are kinetically stable compared to conventional emulsions. Furthermore, they are transparent or translucent to the naked eye [15]. They have also been found to hold special characteristics such as protection from oxidant and hydrolysis in oil-in-water

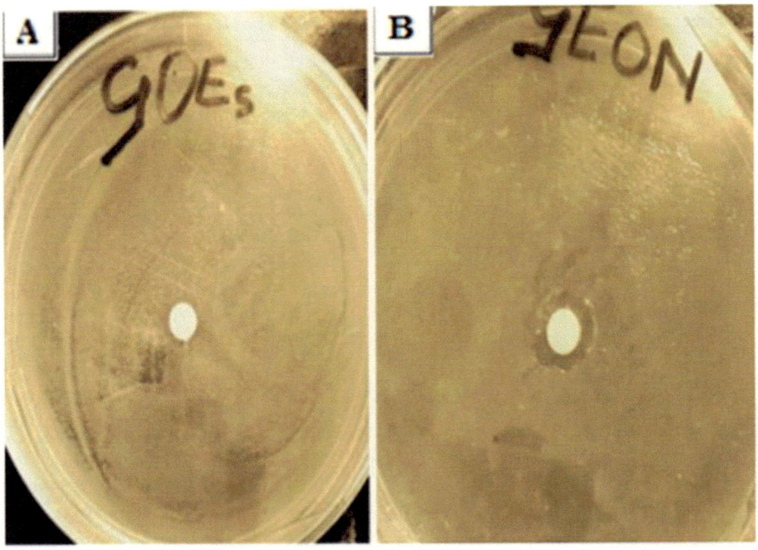

Figure 3.
Photographic evidence of the antimicrobial inhibition of (A) garlic essential oil and (B) garlic essential oil nanoemulsions [19].

(O/W) nanoemulsions [16], encapsulation of hydrophilic drugs [17], enhanced bioavailability of drugs [18], and increased antimicrobial activity of essential oil [19]. **Figure 3** shows an enhanced inhibition level against *Escherichia coli* of garlic essential oil nanoemulsions (GEON) as compared to garlic essential oil [19]. The study by Katata-Seru et al. further revealed an easy and effective Taguchi method for optimizing GEON and as a potential alternative to antimicrobial broiler growth promoters.

There are various types of nanoemulsions and the most common ones are O/W type, water-in-oil (W/O) type, and bi-continuous type, for example, water-in-oil-in-water (W/O/W) type. A number of different preparation techniques for nanoemulsions have been investigated intensively using low and high energy. Low energy includes spontaneous emulsification and phase inversion temperature, while high-energy such as microfluidics, high-pressure homogenizers, or ultrasound equipment methods are used. The food-grade nanoemulsions have generated a huge interest using processing operations such as homogenization and mixing and shearing and homogenization [7, 20]. Recently, Öztürk evaluated various studies on enhanced bioavailability of vitamins A, D, and E encapsulated in O/W nanoemulsions and their factors affecting their stability [16]. Emulsion systems for encapsulation of vitamin E showed that nanoemulsion formulation improved the emulsion stability with an average particle size of 277 nm when compared to the standard emulsion [20]. Although it appears that significant research on nanoemulsions is on the rise, Öztürk highlighted a need for more in vivo bioavailability studies of the foods fortified with lipophilic vitamins as their studies are few owing to the higher costs.

1.2 Polymeric nanoparticles

Polymeric nanoparticles (NPs) are solid carriers capable to adsorb, disperse, entrap, and attach active ingredients to its matrices with the size of smaller than 1 μm. They are produced from preformed polymers by emulsion solvent evaporation, salting out, dialysis, nanoprecipitation, and supercritical fluid (SCF) technology. The NPs have displayed fairly good stability, higher loading efficiency, and controlled release of bioactive compounds as compared to emulsion, micelles, and liposomes [14]. In addition, they have been studied extensively in the nutraceutical field because of characteristics including increased stability, the capability to protect drugs, etc. [21, 22].

Studies have fabricated a unique polymeric vitamin E-modified aliphatic polycarbonate (mPEG-PCC-VE) to assist oral absorption of oleanolic acid (OA) [23]. The OA demonstrated excellent pharmacological activities in the clinical treatment of hypoglycemia, immune regulation, acute jaundice, and chronic toxic hepatitis. In spite of this, OA has limited water solubility and poor intestinal mucosa permeability when delivered orally. The results of OA encapsulated mPEG-PCC-VE NPs illustrated homogeneous 170 nm particle size with a drug loading of 8.9% and a potential platform to facilitate the oral delivery of OA.

1.3 Lipid nanoparticles

Lipid nanoparticles were developed as an alternative to traditional nano-systems such as polymeric particles and liposomes. Lipid nanoparticles can be defined as colloidal particles composed of lipids stabilized by surfactants that are solid at ambient temperature with sizes varying between 40 and 1000 nm [24, 25]. The first lipid nanoparticle to be produced was solid lipid nanoparticles as depicted in **Figure 4**. SLN is made from solid lipid only. The second generation of lipid nanoparticles was developed a few years later called nanostructured lipid nanoparticles (NLC). NLC is made from a blend of solid and liquid (oil) lipids [26]. The addition of oil in NLC formulation is meant to distort the formation of perfectly structured lipid crystals found in SLN, thus creating more room with uptake capacity for the encapsulated active. This was first shown by Jenning and Gohla, when they increased the loading capacity of retinol (vitamin A) from 1 to 5% by using NLC [27].

Lipid particles can be produced using various methods such as high-pressure homogenization, microemulsion, emulsion solvent evaporation, emulsification-solvent diffusion, solvent displacement, phase inversion, ultrasonication, and membrane contractor technique [24]. However, of these techniques, only a few have been applied to prepare lipid nanoparticles with vitamins. Vitamins are sensitive

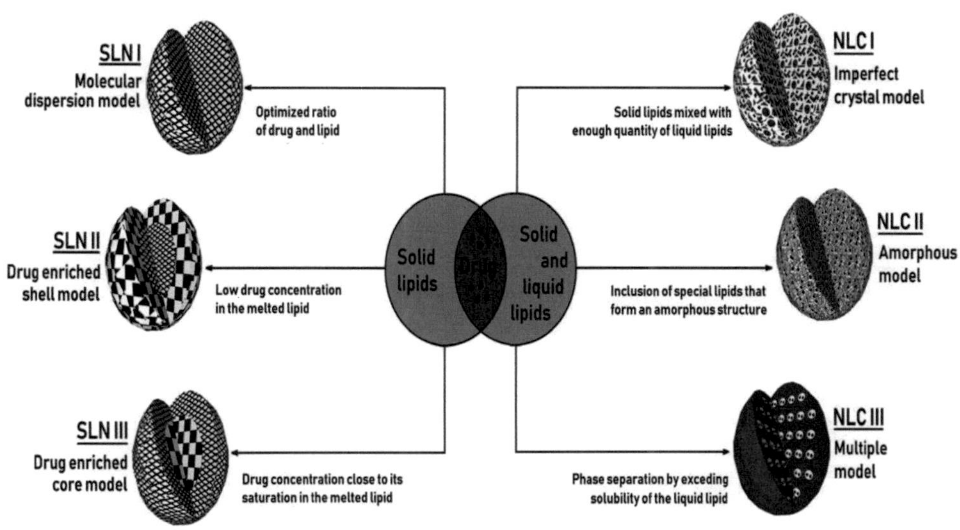

Figure 4.
Models for the structure solid lipid nanoparticles and nanostructured lipid carriers that can be obtained under different conditions determined by the nature of the components and their relative solubility [25].

bioactives, and thus care should be taken to employ techniques that will retain their activity during formulation. The most widely used techniques to prepare lipid nanoparticles with vitamins are the emulsion solvent evaporation method, high-pressure homogenization, and microemulsions.

1.3.1 Emulsion solvent evaporation method

This method is based on the dispersion of a solution of the lipid components in an aqueous surfactant solution. The lipids and the lipophilic bioactive are commonly dissolved in an organic solvent such as dichloromethane, cyclohexane, ethyl acetate, or chloroform. When the nanoemulsion is formed, the solvent is extracted or evaporated, and the droplets start to solidify until solid lipid nanoparticles encompassing the active are formed. The solvent can be evaporated by agitation, rotary evaporation, or spray-drying [28]. The emulsion solvent evaporation method offers a great advantage for encapsulating actives that are highly sensitive to heat such as vitamins as no thermal stress is needed [29].

1.3.2 High-pressure homogenization

This technique involves the preparation of a pre-emulsion, which is then passed under high pressure (100–2000 bar) through a homogenizer valve. The pre-emulsion generally composes a lipid phase and an aqueous phase containing a surfactant. The fluid is accelerated in a very short distance in the homogenizer, reaching a high speed. The lipid substances are then divided into small droplets by the shear stress forces. This technique may produce particles with low encapsulation efficiency for hydrophilic substances due to the drug migrating to the external aqueous phase during particle formation. However, lipophilic actives can be encapsulated at high dosage [30, 31]. The technique is also ideal for the production of large quantities of sterile particles, which is an advantage for nutraceuticals [26].

1.3.3 Microemulsions

Microemulsions are clear, thermodynamically stable, and isotropic liquid mixtures of oil, water, and surfactant and almost always co-surfactant as well. The droplet size in the dispersed phase of the microemulsion is less than 100 nm. The droplets are formed by the drastic cooling of a microemulsion mixture to solidify the droplets and create particles loaded with the bioactive. The preparation of microemulsions does not require much energy to form and is thus recommended for actives that are highly sensitive to shear forces or thermal stress as is the case with most vitamins [29]. Other advantages of microemulsions include the use of bioactive compatible ingredients and the enhanced stability of formulations, as they are thermodynamically stable [32].

After preparation, lipid nanoparticles can be stored as nanosuspensions in the medium they were formed, or dry particles can be obtained using either freeze-drying or spray-drying [25]. In some instances, aggregation of particles may occur due to the drying process. In these instances, an adequate amount of cryoprotectant can be added to prevent or minimize aggregation of the particles [30].

1.3.4 Lipid nanoparticles for the delivery of vitamins

There has been an increasing awareness of maintaining personal health by balanced nutrition and the intake of nutraceutical supplements. Due to the challenges faced with the stability of nutraceuticals, lipid formulations have been sought to enhance the stability of these bioactives. Lipid nanoparticles are highly recommended for the delivery of vitamins firstly due to their physical stability, secondly due to their ability to protect actives from environmental factors such as oxidation, hydrolysis, and possibly enzymatic degradation in the gastrointestinal tract, and thirdly due to their cost-effective production at large scale, for example, high-pressure homogenization.

Following the study by Jenning and Gohla, liposoluble vitamins A, D, E, and K as well as their derivatives have been presented as good candidates for encapsulation using lipid systems due to their low bioavailability and relative instability [33]. With the aim of increasing the intestinal absorption of vitamin K, vitamin K1 was encapsulated in SLN and demonstrated stability for more than 2 days in simulated gastric and intestinal fluids. The encapsulation also increased storage stability of vitamin K1 up to 4 months at 25°C [34]. Vitamin A in the form of all trans-retinol suffers degradation reactions that are characteristic of conjugated double bonds resulting in loss of its bioactivity. The vitamin was encapsulated in SLN and demonstrated an enhancement of retinol stability, photostability, and preservation of its antioxidant activity [35]. The lipophilicity, chemical instability, and poor skin penetration of vitamin E have limit its effectiveness as an antioxidant and photoprotectant used in various pharmaceutical and cosmetic products [25]. Various lipid formulations with vitamin E were developed, and the results obtained showed the possibility to enhance chemical stability and physical stability of Vitamin E in a cream [36]. In other formulations, Tween 80 was mixed with various lipids and surfactants to produce particles with the ability to protect vitamin E against photodegradation [37, 38]. Ying and Misran produced a thermoresponsive gel for topical application that could control the release of vitamin E [39].

1.4 Polymeric micelles

Polymeric micelles are formed from block copolymers that have amphiphilic character. Amphiphilic polymers are copolymers composed of hydrophilic ("water-loving") and hydrophobic ("water-hating") parts [40]. They normally form spontaneously under certain concentrations and temperatures in a given media [41]. The concentration at which these micelles are formed is known as critical micelle concentration (CMC), while the temperature at which this micelle exists is called the critical micellization temperature (CMT) [41]. Hydrophobic blocks of amphiphilic polymers form the core of the micelle, while the hydrophilic blocks form the shell [42, 43]. They can be utilized as drug carriers, by incorporating the poorly soluble nonpolar substances within the micellar core and polar substances on the micellar shell (by adsorption); substances with intermediate polarity are distributed between the core and shell [41, 44]. These properties enable these systems to incorporate poorly water-soluble drugs in the micellar core by physical interaction or by chemical conjugation leading to higher solubility extents [43], to protect the drugs or sensitive substances from premature degradation and also reduce the toxicity of the drug [42]. When compared to conventional micelles, polymeric micelles have lower CMCs values and are more stable even at concentrations below CMC [41]. This behavior stems from the slower rate of dissociation that depends on the molecular weight and hydrophilic-hydrophobic balance of the polymer as well as the properties of the drug incorporated into the core [43].

1.5 Supercritical fluid technology

The methods discussed in the preceding sections involve the use of organic solvents, which could impart residual moisture of organic solvents on the produced nanoparticles. Supercritical fluid technology, on the other hand, utilizes the CO_2, which often produces nanoparticles or microparticles without any trace of solvent, thus high purity. CO_2 is a cheaper fluid, nontoxic, and non-flammable. Its low critical temperature of 31.1°C makes it an ideal fluid for sensitive or thermally labile materials. An active ingredient is regarded to be in a supercritical state if its temperature and pressure are above its critical values. Based on the solubility of active ingredient in CO_2 (or any inert gas) fluid, particles can be formed by using two approaches as depicted in **Figure 5**: (a) rapid expansion of supercritical solution (RESS) and (b) rapid expansion of supercritical solution into a liquid solvent (RESOLV) [45]. In order to perform RESS, high solubility in the supercritical fluid is required. However, some of the active nutraceutical ingredients are organic polar compounds. CO_2, due to its low polarity, is not a proper fluid for these materials. The nutraceutical ingredients are ideal for RESOLV. In RESOLV, an organic solvent is required to dissolve the vitamins expanding in the SCF.

As discussed in Section 2.3, solid lipid nanoparticles are spherical nanoparticles produced from solid fat. Instead of melting the lipids in an appropriate organic solvent, vitamin B2 was encapsulated in SNL using SCF [46]. The lipids are saturated with CO_2 in order to decrease the melting point. However, Couto and colleagues modified the SCF, in which the lipid, bioactive, and surfactant mix expanded with CO_2 was decompressed into a water stream containing a stabilizer. Vitamin B2, the hydrophilic bioactive, was encapsulated in fully hydrogenated canola oil (the solid lipid), using sodium lauryl sulfate as surfactant and polyethylene glycol as stabilizer. Vitamin B2 participates in a range of redox reactions central to human metabolism, and its deficiency has been linked to fetal developmental abnormalities and deficiencies in the production of red blood cells. Due to its hydrophilic nature, it is easily absorbed, but it is not stored in the body, leading to the need for

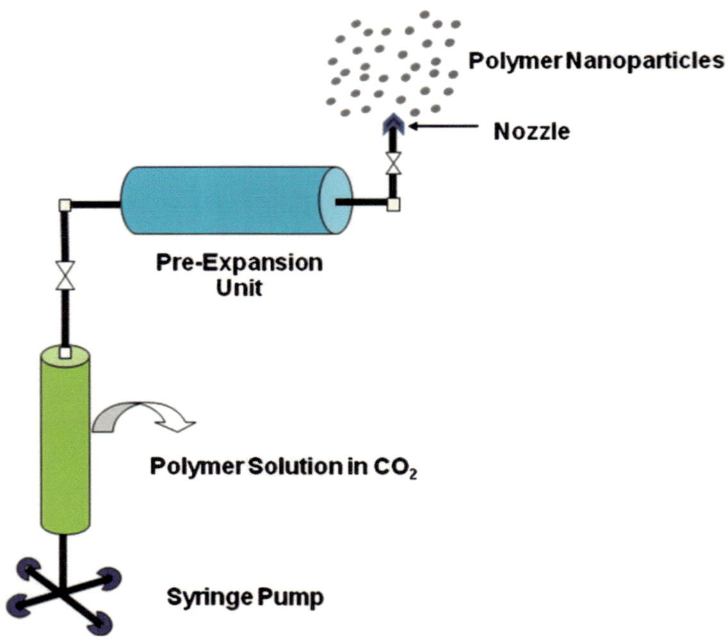

Figure 5.
A Schematic representation of particle formation by rapid expansion of supercritical solution. In conventional RESS, blank particles are formed dissolving the solute, that is, polymer in a supercritical fluid to form a solution. This is followed by the rapid expansion of the solution across an orifice or a capillary nozzle into ambient air. The high degree of supersaturation, accompanied by the rapid pressure reduction in the expansion unit, ideally results in homogenous growth of particles and, thereby, the formation of well-dispersed particles. However, results obtained from mechanistic studies of different model solutes for the RESS process indicate that both nanometer- and micrometer-sized particles are usually present in the expansion unit [46].

replenishing its levels every day. By encapsulating it in SLN, it is anticipated that a sustained release can be obtained and its absorption can be slowed down.

2. Conclusions

Numerous delivery systems such as nanoemulsions, microemulsions, liposomes, lipid nanoparticles, polymeric micelles, and nanoparticles have been reported extensively for encapsulating nutraceuticals especially liposoluble vitamins. The lipid-based nanoformulations were highly recommended by various studies as compared to others due to their better absorption when ingested, improved stability, and low degradation in the gastrointestinal tract. However, more efforts need to be focused on their toxicity and regulatory issues for the faster development of industrially processed nanoformulated nutraceuticals from the lab-scale research discoveries. It is widely anticipated that over the next couple of years, nanoformulated delivery systems of essential nutraceuticals will continue to evolve and many novel food products are expected to be used with an enormous positive impact on addressing malnutrition challenges in children.

Author details

Lebogang Katata-Seru[1]*, Bathabile Ramalapa[2] and Lesego Tshweu[2]

1 Department of Chemistry, School of Physical and Chemical Sciences, North-West University, Mmabatho, South Africa

2 Polymer and Composites, CSIR Materials Science and Manufacturing, Pretoria, South Africa

*Address all correspondence to: lebo.seru@nwu.ac.za

References

[1] DeFelice SL. What is a true nutraceutical? And what is the nature and size of the U. S. Market? Available from: http://www.fimdefelice.org/p2504.html [Accessed: December 13, 2018]

[2] Das L, Bhaumik E, Raychaudhuri U, Chakraborty R. Role of nutraceuticals in human health. Journal of Food Science and Technology. 2012;**49**(2):173-183

[3] Dutta S, Ali KM, Dash SK, Giri B. Role of nutraceuticals on health promotion and disease prevention: A review. Journal of Drug Delivery and Therapeutics. 2018;**8**(4):42-47

[4] Gonçalves RFS, Martins JT, Duarte CMM, Vicente AA, Pinheiro AC. Advances in nutraceutical delivery systems: From formulation design for bioavailability enhancement to efficacy and safety evaluation. Trends in Food Science & Technology. 2018;**78**:270-291

[5] Chawda PJ, Shi J, Xue S, Young Quek S. Co-encapsulation of bioactives for food applications. Food Quality and Safety. 2017;**1**(4):302-309

[6] Chauhan B, Kumar G, Kalam N, Ansari SH. Current concepts and prospects of herbal nutraceutical: A review. Journal of Advanced Pharmaceutical Technology & Research. 2013;**4**(1):4-8

[7] Aditya NP, Espinosa YG, Norton IT. Encapsulation systems for the delivery of hydrophilic nutraceuticals: Food application. Biotechnology Advances. 2017;**35**:450-457

[8] Giroux HJ, Constantineau S, Fustier P, Champagne CP, St-Gelais D, Lacroix M, et al. Cheese fortification using water-in-oil-in-water double emulsions as carrier for water soluble nutrients. International Dairy Journal. 2013;**29**(2):107-114

[9] Alishahi A, Mirvaghefi A, Tehrani MR, Farahmand H, Shojaosadati SA. Shelf life and delivery enhancement of vitamin C using chitosan nanoparticles. Food Chemistry. 2011;**126**:935-940

[10] de Brittoa D, de Mourab MR, Aouadac FA, Luiz HC, Mattosoa LHC, Assis OBG. N,N,N-trimethyl chitosan nanoparticles as a vitamin carrier system. Food Hydrocolloids. 2012;**27**:487-493

[11] Ravisankar P, Reddy AA, Nagalakshmi B. The comprehensive review on fat soluble vitamins. IOSR Journal of Pharmacy. 2015;**5**(11):12-28

[12] http://www.faqs.org/nutrition/Smi-Z/Vitamins-Fat-Soluble.html [Accessed: January 10, 2019]

[13] https://www.healthline.com/nutrition/fat-soluble-vitamins#vite [Accessed: January 10, 2019]

[14] Shin GH, Kim JT, Park HJ. Recent developments in nanoformulations of lipophilic functional foods. Trends in Food Science & Technology. 2015;**46**:144-157

[15] Solans C, Izquierdo P, Nolla J, Azemar N, Garciacelma M. Nano-emulsions. Current Opinion in Colloid & Interface Science. 2005;**10**(3-4):102-110

[16] Ozturk B. Nanoemulsions for food fortification with lipophilic vitamins: Production challenges, stability, and bioavailability. European Journal of Lipid Science and Technology. 2017;**119**:1-18

[17] Tshweu L, Katata L, Kalombo L, Swai H. Nanoencapsulation of water-soluble drug, lamivudine, using a double emulsion spray-drying technique for improving HIV treatment. Journal of Nanoparticle Research. 2013;**15**:2040, 1-11

[18] Tshweu L, Katata L, Kalombo L, Chiappetta DA, Hocht C, Sosnik A, et al. Enhanced oral bioavailability of the antiretroviral efavirenz encapsulated in poly(epsilon-caprolactone) nanoparticles by a spray-drying method. Nanomedicine (Lond). 2014;**9**(12):1821-1833

[19] Katata-Seru L, Lebepe TC, Aremu OS, Bahadur I. Application of Taguchi method to optimize garlic essential oil nanoemulsions. Journal of Molecular Liquids. 2017;**244**:279-284

[20] Hategekimana J, Chamba MVM, Shoemaker CF, Majeed H, Zhong F. Vitamin E nanoemulsions by emulsion phase inversion: Effect of environmental stress and long-term storage on stability and degradation in different carrier oil types. Colloids and Surfaces A: Physicochemical and Engineering Aspects. 2015;**483**:70-80

[21] Man DK, Casettari L, Cespi M, et al. Oleanolic acid loaded PEGylated PLA and PLGA nanoparticles with enhanced cytotoxic activity against cancer cells. Molecular Pharmaceutics. 2015;**12**:2112-2125

[22] Gonnet M, Lethuaut L, Boury F. New trends in encapsulation of liposoluble vitamins. Journal of Controlled Release. 2010;**146**:276-290

[23] Zhang W, Liang C, Liu H, Li Z, Chen R, Zhou M, et al. Polymeric nanoparticles developed by vitamin E-modified aliphatic polycarbonate polymer to promote oral absorption of oleanolic acid. Asian Journal of Pharmaceutical Sciences. 2017;**12**:586-593

[24] Shah M. Solid lipid nanoparticles (SLN) for oral drug delivery: An overview. Journal of Nanoscience and Nanomedicine. 2017;**2017**:1-2

[25] Saez V, Souza D, Mansur R. Lipid nanoparticles (SLN & NLC) for delivery of vitamin E: A comprehensive review. Journal of Cosmetic Science. 2018;**40**:103-116

[26] Muller R, Shegokar R, Keek C. 20 years of lipid nanoparticles (SLN & NLC): Present state of development and industrial applications. Current Drug Discovery Technologies. 2011;**8**:207-227

[27] Jenning V, Gohla S. Encapsulation of retinoids in solid lipid nanoparticles (SLN). Journal of Microencapsulation. 2011;**18**:149-158

[28] Iqbal MA, Md S, Sahni JK, Baboota S, Dang S, Ali J. Nanostructured lipid carriers system: Recent advances in drug delivery. Journal of Drug Targeting. 2012;**20**:813-830

[29] Battaglia L, Gallarate M. Lipid nanoparticles: State of the art, new preparation methods and challenges in drug delivery. Expert Opinion on Drug Delivery. 2012;**9**:497-508

[30] Ekambaram P, Sathali AAH, Priyanka K. Solid lipid nanoparticles: A review. Scientific Reviews and Chemical Communications. 2012;**2**:80-102

[31] Mehnert W, Mäder K. Solid lipid nanoparticles: Production, characterization and applications. Advanced Drug Delivery Reviews. 2012;**64**:83-101

[32] Gupta DR, Shah YD, Vora RS, Shah D. Solubility enhancement by solid lipid nanoparticle. IJPPR Human. 2016;**7**:351-367

[33] Fangueiro JF, Macedo AS, Jose S, Garcia ML, Souto SB, Souto EB. Thermodynamic behavior of lipid nanoparticles upon delivery of vitamin E derivatives into the skin: In vitro studies. Journal of Thermal Analysis and Calorimetry. 2012;**108**:275-282

[34] Liu CH, Wu CT, Fang JY. Characterization and formulation

optimization of solid lipid nanoparticles in vitamin K1 delivery. Drug Development and Industrial Pharmacy. 2010;**36**:751-761

[35] Jee J-P, Lim S-J, Park J-S, Kim C-K. Stabilization of all-trans retinol by loading lipophilic antioxidants in solid lipid nanoparticles. European Journal of Pharmaceutics and Biopharmaceutics. 2006;**63**:134-139

[36] Dingler A, Blum R, Niehus H, Muller R, Gohla S. Solid lipid nanoparticles (SLNTM/LipopearlsTM) a pharmaceutical and cosmetic carrier for the application of vitamin E in dermal products. Journal of Microencapsulation. 1999;**16**:751-767

[37] Abla M, Banga A. Formulation of tocopherol nanocarriers and in vitro delivery into human skin. International Journal of Cosmetic Science. 2014;**36**:239-246

[38] Chen J, Wei N, Lopez-Garcia M, Ambrose D, Lee J, Annelin C, et al. Development and evaluation of resveratrol, Vitamin E, and epigallocatechin gallate loaded lipid nanoparticles for skin care applications. European Journal of Pharmaceutics and Biopharmaceutics. 2017;**117**:286-291

[39] Ying LQ, Misran M. Rheological a physicochemical characterization of alphatocopherol loaded lipid nanoparticles in thermoresponsive gel for topical application. Malaysian Journal of Fundamental and Applied Sciences. 2017;**13**:248-252

[40] Anderson PM, Wilson MR. Molecular dynamics simulations of an amphiphilic graft copolymer at a water/air interface. The Journal of Chemical Physics. 2004;**121**(17):8503-8510

[41] Torchilin VP. Targeted polymeric micelles for delivery of poorly soluble drugs. Cellular and Molecular Life Sciences. 2004;**61**:2549-2559

[42] Kataoka K, Harada A, Nagasaki Y. Block copolymer micelles for drug delivery: Design, characterization and biological significance. Advanced Drug Delivery Reviews. 2001;**47**:113-131

[43] Chiappetta DA, Alvarez-Lorenzo C, Rey-Rico A, Taboada P, Concheiro A, Sosnik A. N-alkylation of poloxamines modulates micellar assembly and encapsulation and release of the antiretroviral efavirenz. European Journal of Pharmaceutics and Biopharmaceutics. 2010;**76**(1):24-37

[44] Sosnik A, Carcaboso AM, Glisoni RJ, Moretton MA, Chiappetta DA. New old challenges in tuberculosis: Potentially effective nanotechnologies in drug delivery. Advanced Drug Delivery Reviews. 2010;**62**(4-5):547-559

[45] Nagavarma BVN, Hemant KS, Yadav AA, Vasudha LS, Shivakumar HG. Different techniques for preparation of polymeric nanoparticles. Asian Journal of Pharmaceutical and Clinical Research. 2012;**5**(3):16-23

[46] Rao JP, Geckeler KE. Polymer nanoparticles: Preparation techniques and size-control parameters. Progress in Polymer Science. 2011;**36**(7):887-913

Section 3

Fabrications

Chapter 4

Development of Nano-Emulsions of Essential Citrus Oil Stabilized with Mesquite Gum

Maira Berenice Moreno-Trejo,
Arturo Adrián Rodríguez-Rodríguez, Ángela Suarez-Jacobo and Margarita Sánchez-Domínguez

Abstract

The use of nano-emulsions has great advantages over conventional macro-emulsions since the small droplet size allows to expand the options of applications besides presenting a greater surface area. This chapter focuses on the formulation of nano-emulsions of citrus essential oils in water, stabilized with a natural gum (mesquite gum), using a high pressure microfluidic homogenizer to obtain appropriate physicochemical characteristics and kinetic stability. When establishing the general conditions of the methods for obtaining nano-emulsions by high pressure homogenization, several formulations presented stability and size corresponding to nano-emulsions, and these were monitored during 4 months in order to study their stability as a function of time. Taking into account the results of size and stability, the best nano-emulsion obtained had a composition of Persian lemon oil (9.86%), mesquite gum (4.93%) Tween 80 (4.89%), Span 20 (1.45%), and deionized water (78.86%) with an average droplet size of 40 nm. In addition, the antibacterial activity studies also showed that this formulation had the best performance against common bacteria such as *Staphylococcus aureus* and *Escherichia coli*. The analysis of the minimum inhibitory concentration (MIC) shows that it is possible to prevent the growth of these particular bacteria using 6.25% of the best nano-emulsion formulations.

Keywords: nano-emulsions, essential oil, mesquite gum, high pressure microfluidic homogenizer, antibacterial

1. Introduction

1.1 Nano-emulsions

Nano-emulsions can be defined as ternary systems composed of nano-sized droplets (20–500 nm) dispersed in a continuous phase and stabilized by amphiphilic molecules known as surfactants [1–4]. The nature of the droplets and continuous phase can be aqueous or oily, forming oil-in-water (O/W) or water-in-oil (W/O) nano-emulsions. Regarding surfactants, these molecules must possess a double affinity, a hydrophilic polar head and a hydrophobic non-polar tail, for example. The surfactant allows decreasing the oil/water interfacial tension (e.g. ~5–10 mNm^{-1}), leading to formation and dispersion of very small droplets. The decrease in interfacial tension can be enhanced by the addition of a co-surfactant, which acts in similar way than surfactants, but is added in a much lower content. The singularity of nano-emulsions is its small droplet size range, kinetic stability, and the lack of surfactant (5–15% is usually needed) compared to other emulsion systems such as microemulsions [5, 6]; but, in order to be formed they require an external stimulus (being kinetically stabilized) [7] as:

a. Mechanical input: applied with a special equipment such as ultrasonic probe (UP) or high pressure homogenizer (HPH). In a typical HPH (**Figure 1**), a pump pushes a pre-formulated macroemulsion through a narrow gap, a microfluidic interaction chamber (in the micrometer range), where the large droplets break into smaller droplets due the extreme elongation and shear stresses through the homogenization process. This step can be repeated multiple times until the droplet size becomes constant. On the other hand, the UP can induce the breaking of pre-formulated macroemulsion droplets by cavitation, which can also be repeated numerous times. The size and polydispersity of these droplets depends of the applied energy through the whole nano-emulsification process, as well to the water/oil/surfactant ratio [1, 5].

b. Physicochemical input: requires a change of temperature (PIT: phase inversion temperature method) or the addition of an extra amount of dispersed phase (PIC: phase inversion composition). In the first case, the increasing or decreasing of temperature can induce a phase transition (e.g. O/W ➔ W/O), due the change of the hydrophilic–lipophilic balance (HLB) of the system. In the second case, the increment of the droplet constituent can lead to the dispersed-phase/continuous phase ➔ continuous-phase/dispersed phase transition [8, 9]. Both approaches are based on phase changes that pass through a zero-curvature system such as a bicontinuous microemulsion or a lamellar phase, in order to facilitate phase inversion [3].

Compared with mechanical methods, physicochemical strategies are considered low-energy methods (~103 Wkg^{-1} vs. ~1010 Wkg^{-1}); however, they need

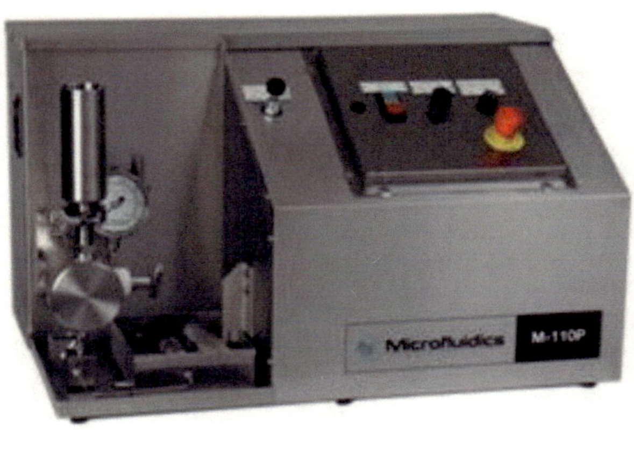

Figure 1.
Typical high pressure homogenizer equipment.

very detailed phase behavior studies, which are time consuming. For this reason, high-energy methods may be preferred. Regarding applications, nano-emulsions can be employed in a wide range of fields such [10]: pharmaceutical, cosmetic, automotive and food industry, depending on their physicochemical properties. In this study, citric oil droplets were formed and dispersed in an aqueous continuous phase, using polysorbates as surfactants and a natural gum (mesquite), through an assisted HPH method. The chemistry of the citric oil and nano-emulsion stability allows the application of the formulated systems for antibacterial applications against bacteria such as *Escherichia coli* and *lactobacillus delbrueckii* at studied conditions.

1.2 Polysorbate surfactants

Tween 80 and Span 20 (**Figure 2**) belong to the polysorbate family, a non-ionic type of surfactants. Chemically, polysorbates are derived from ethoxylated sorbitan, a derivative of sorbitol, which is esterified with fatty acids. Tween and Span are proprietary names from CRODA™ (manufacturer of specialty chemicals); the numeric values are specific to the fatty acid derivative: oleic acid, for Tween 80, and lauric acid, for Span 20. Both surfactants are frequently used as emulsifiers for the food and cosmetic industry, having a very low toxicity and eco-friendly chemistry [11–13]. However, the affinity for polar (water) and non-polar (oil) groups is different for each non-ionic surfactant; according to the hydrophilic–lipophilic balance scale (HLB) [14], Tween 80 is hydrophilic (HLB: 15), while Span 20 is more lipophilic (HLB: 8.6).

1.3 Mesquite gum as emulsifier

Mesquite gum is a vitreous exude, produced by mesquite tree (*Prosopis laevigata*), a widely distributed plant across arid and semiarid zones. This gum is composed of a highly tailored polysaccharide salt, constituted by residues of L-arabinose, D-galactose, 4-O-Methyl-D-glucuronic acid and L-rhamnose (**Figure 3**) [15, 16]. Mesquite gum chemical composition is similar (different molar ratio) to that one of Arabic gum, which is commonly employed in the food and pharmaceutical industry, due its emulsification capacity [17, 18]. In México, mesquite gum is only consumed as a candy, therefore there is a wield field for exploration of this product as emulsifier.

1.4 Citric oil nano-emulsions

A natural antibacterial extracted from plant or fruit origin is the essential oil, many studies have described this effect [19–21]. Pink Grapefruit

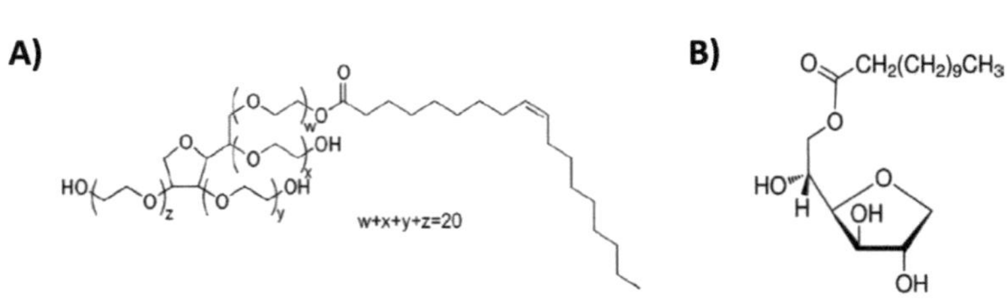

Figure 2.
(A) Tween 80 and (B) Span 20 chemical structures.

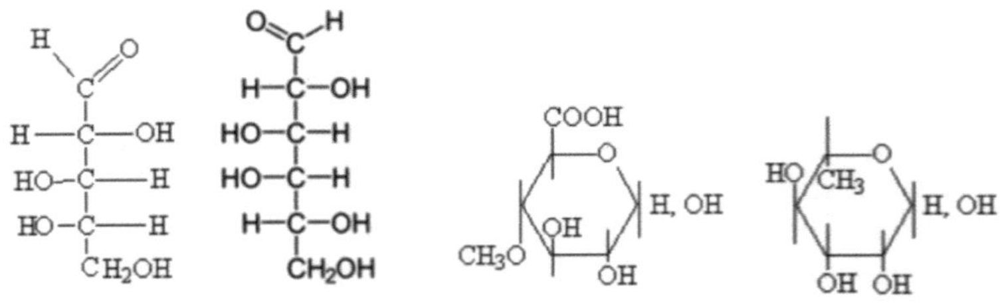

Figure 3.
Main constituents of mesquite gum (from left to right): L-arabinose, D-galactose, 4-O-Methyl-D-glucuronic acid and L-rhamnose.

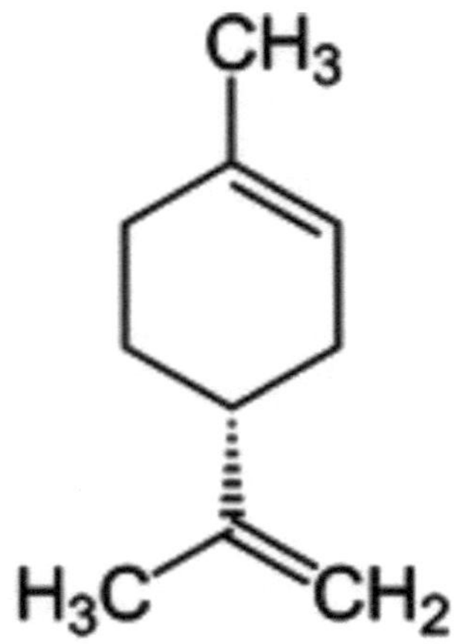

Figure 4.
D-limonene chemical structure.

(*Citrus paradise*) belongs to the Citrus genus, a class of flowering plants in the family Rutaceae [22]. The active constituents exist in this kind of citrus essential oil, such as limonene, a-pinene, b-pinene and a-terpinolene [22, 23]. Its seed and peel extract (essential oil) has shown the antibacterial and antifungal property [24]. Persian lemon (*citrus latifolia tanaka)* is composed of citric, malic and formic acids, as well from pectin, hesperidin, and essential oils as D-limonene (**Figure 4**) and phellandrene, which are the volatile liquid fractions responsible for the characteristic smell of lemon. These compounds can be extracted by distillation, and are commonly employed in the cosmetic, pharmaceutical and food industry [25].

As part of a nano-emulsion, D-limonene has been used as the oil phase of different O/W nano-emulsions. For instance, in the work of Li and Chiang, who have successfully formed D-limonene O/W nano-emulsions phase by an ultra-sonic method, using Span 85 and Tween 20 as surfactants [26]. Another example is the work of Donsì et al. [27], who achieved D-limonene nano-emulsions formation through a high pressure homogenizer, using a wide type of emulsifiers. The latter authors have explored the antibacterial properties of these systems by testing against *Escherichia coli* and *lactobacillus delbrueckii*, demonstrating D-limonene nano-emulsion capacity to control and eliminate microbial organisms [27, 28]. As these studies, there are some other investigations about D-limonene nano-emulsions, but none of such studies explores the formulation with a natural emulsifier as mesquite gum.

2. Analysis of the components of a nano-emulsion

The first stage of the research consists of the characterization of the raw materials (gums and citric essential oils) used in the formulation of nano-emulsions.

2.1 Mesquite gum purification

Mesquite gum was extracted from mesquite pearls obtained from a local candy store at Sonora (Sonora, México). The purification process of the mesquite gum begins with the selection of the cleanest pearls by visual inspection as indicated by literature [17]; afterwards the pearls were ground in a ceramic mortar until a powder was obtained. Then a 20% (wt) solution of this powder was prepared, which was filtered (in order to eliminate impurities like dust and pieces of wood), details of the process can be found in literature [12].

Afterwards, several drying processes were experimented, specifically oven and lyophilization. The drying process by oven was discarded since it was very slow, and in addition, the gum obtained was of a brown tone, indicating a possible degradation of the gum or caramelization. Regarding the lyophilization method, several strategies were tested. First, it was tested with a lyophilized sample at 20%, and a gum with a "sponge" texture was obtained, later this "sponge" gum was re-lyophilized in order to obtain a denser version. Therefore, it was decided to test lyophilized 40% mesquite gum solutions in a controlled manner, this last process was selected to be used for the production of gum for the nano-emulsions of this research.

2.2 Infrared spectrum analysis (FT-IR) of the gums

To carry out this analysis, the Thermo Nicolet 6700 FT-IR spectrophotometer with its attenuated total reflectance accessory (Thermo Scientific) was used, using solid samples (powders).

Figure 5 shows the FT-IR spectra of the different gums characterized in this research, which are largely comparable with the FT-IR spectrum of Arabic gum (GA) reported in the literature [17, 18]. Therefore, it follows that the different mesquite gums produced (oven dried procedure, dense lyophilized and controlled Lyophilized) essentially have the same chemical composition in terms of sugars, amino acids and proteins. Absorbance of the —OH and —CH groups are observed at 3375 and 2932 cm^{-1} (similar to that reported in the literature [17]), and a band centered between 1650 and 1600 cm^{-1} that can be assigned to the primary amides. There is a smaller band around 1500 cm^{-1} that is assigned to the secondary or substituted amides. The bands of the primary and secondary amides are characteristic of the presence of peptide bonds and confirm the presence of the protein [17], There is also a band at 1400 cm^{-1} that can be attributed to a carboxylic group. The bands that are around 1000 and 900 cm^{-1} can be attributed to the glycosidic acetal groups of pyranose, according to the literature [17].

2.3 Concentration of aldehydes of citric essential oils

The citrus essential oils used during this research were a generous donation from the FRUTECH company. The determination is made by the hydroxylamine hydrochloride method (ISO 1279: 1973). A Titroline Alpha plus automatic titrator from SI Analytics was used to obtain the aldehyde concentration of the citrus oils of pink grapefruit and Persian lemon. Four grams of oil are weighed in a beaker, 50 mL of hydroxylamine hydrochloride solution are added, stirring for 1 minute at 300 rpm, and then it was allowed to rest for 30 minutes. Subsequently, a titration

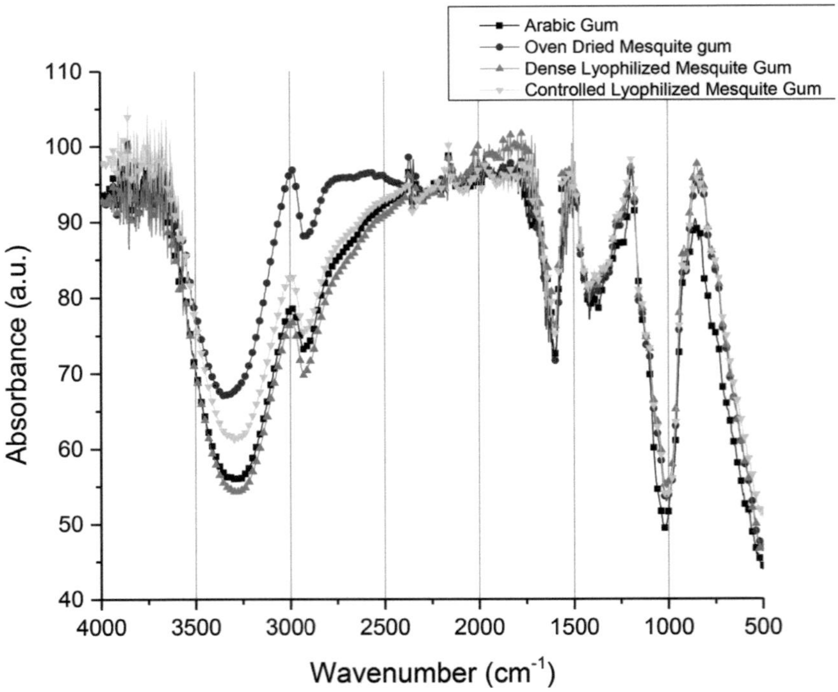

Figure 5.
IR spectra of the different gums.

with methanolic KOH is performed and the amount of mL used is recorded to reach a pH of 3. In order to determine the concentration of aldehydes, the following mathematical formula is used:

$$\%carbonyl\ compounds = (a \times N\,meq \times 100)/p$$

where

a = Volume of the potassium hydroxide solution used in the neutralization of the sample in mL.

N = Normality of the potassium hydroxide solution.

P = Weight of the sample, in grams.

meq = Milliequivalent corresponding to the carbonyl compound in which the result is expressed.

Aldehydes are a family of organic compounds (R—CHO), which are indicative of the quality of essential oils, the higher the concentration of aldehydes, the higher the oil quality [29]. The released HCl is evaluated, which is related to the content of carbonyl groups in the sample and it can be calculated in grams of the aldehyde. The results are shown in **Table 1**, these are within the expected ranges (For Persian lemon is 3.5–7.5 and 0.8 1.5% for pink grapefruit) according to literature [28].

Citric essential oil	%pH	mL	%Aldehydes
Persian lemon	3.89	9.96	3.67
Pink grapefruit	1.07	2.75	1.04

Table 1.
Concentration of Aldehydes.

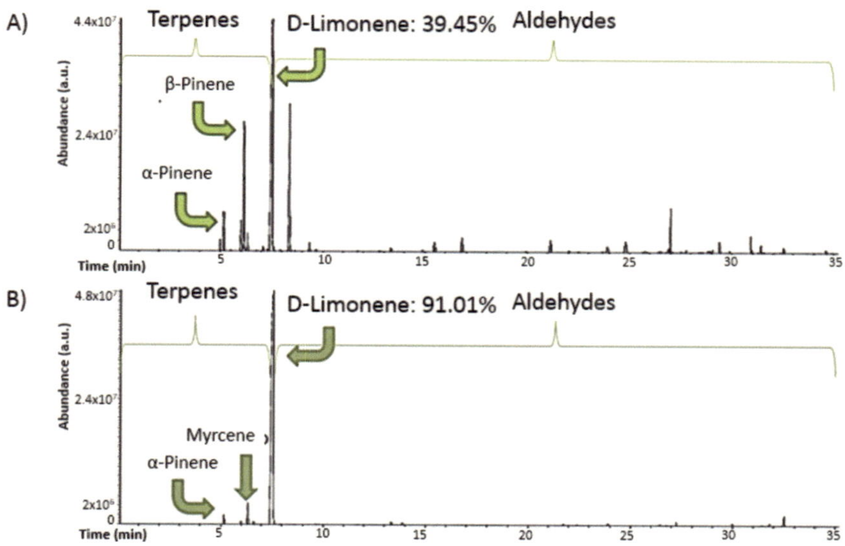

Figure 6.
Chromatographic profile of the sample analyzed: (A) Persian lemon oil and (B) pink grapefruit oil.

It can be seen that Persian Lemon oil has a concentration of aldehydes 3 times greater than pink grapefruit oil, which is why it was a determining factor for the selection of Persian lemon oil as an oil phase. It should be noted that aldehydes are the components with the highest added value since they provide most of the odor, taste and therapeutic qualities, and are therefore the most important components. In addition, some authors attribute the antibacterial activity to aldehydes, although there is controversy in this aspect since other authors give this property to terpenes [30–32].

2.4 Volatile profile by gas chromatography of citric essential oils

The analysis was carried out in a gas chromatograph model 7890ª coupled to a Mass Spectrometer (Agilent Technologies). For the analysis 2 μL of each oil sample is injected, the column HP-5MS (Agilent Technologies) is responsible for passing or retaining each compound of citrus essential oils and uses Helium as a carrier gas at a temperature of 280°C and the Wiley library is used as a database for the identification of each component. This analysis allows the separation and identification of the components of the essential oils. Terpenes have a lower retention time than aldehydes, so this method is used to corroborate the concentration of aldehydes.

As observed in the chromatograms (**Figure 6**) D-limonene is the signal with more abundance in the citrus essential oils used in this research and is a component widely used in microbial inhibition in the food, pharmaceutical and cosmetic industry [33, 34].

The mechanism of the antibacterial function of essential oils is still not detailed, according to the literature [35]. As mentioned earlier there is a controversy regarding antibacterial activity and the relationship with aldehydes and terpenes. As we can see in **Figure 6**, Persian lemon oil has a higher concentration of aldehydes.

3. Methodology for the formulation of nano-emulsions of citrus essential oils

A reproducible process for the formulation of nano-emulsions of essential citrus oils is described. The initial process of preparing the nano-emulsions consists

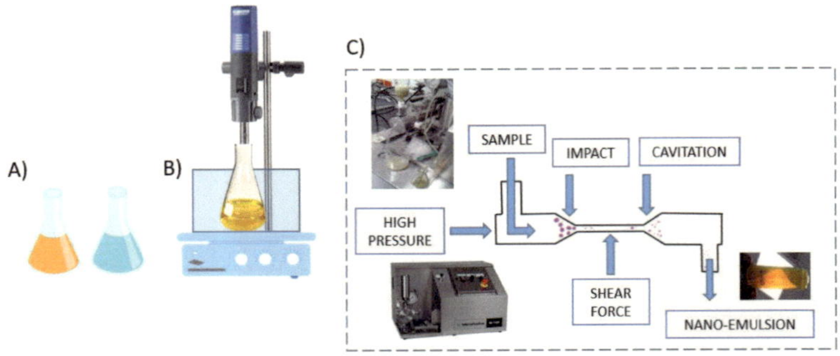

Figure 7.
Process for the formulation of nano-emulsions (A) oil and water phase of nano-emulsions, (B) preparation of pre emulsion by high speed agitation, (C) high pressure process.

of mixing the components in an Erlenmeyer flask using magnetic stirring in a water bath at 40°C for 10 minutes. Followed by high speed mechanical agitation (8000 rpm) for 5 minutes. This first step helps to dissolve the gums and/or surfactants in the aqueous phase as much as possible, and thus it eliminates any possible lumps that could lead to plugging in the microfluidizer interaction chamber. In addition, this step allows to form a pre-emulsion, that is, the oil phase is dispersed as droplets in the aqueous continuous phase; however, at this point, the droplets are micrometric in size and therefore the pre-emulsions have a milky appearance.

This pre-emulsion is then introduced into the high pressure homogenizer (microfluidizer), and subjected to high pressure (the nano-emulsions were subjected to pressures ranging from 10,000 to 30,000 psi) collecting a sample every 1, 3, 5 and 10 laps. In this high pressure process, the droplet size of the nano-emulsion decreases as the number of times the nano-emulsion is introduced to the equipment increases (number of turns or laps or steps), although sometimes the droplet size increases again when the number of laps increases up to a certain value, due to degradation of gums or surfactants. Finally, the nano-emulsion sample is collected and prepared to be sterilized. In **Figure** 7, a summary of the process is presented.

After a series of experiments varying concentrations, applied pressure and number of laps, it was possible to obtain a visually appropriate formulation with relative stability, verified by the characterizations. Therefore, comparison controls are generated, which are described in **Table 2**, replacing mesquite gum with Arabic gum, Tween 80 and Span 20 surfactants, and finally the substitution of mesquite gum for deionized water. These controls were defined in this way to investigate if mesquite gum would have an influence on the characteristics and kinetic stability of the nano-emulsion. Samples of these controls were taken at 1, 3, 5 and 10 steps, **Table 2** shows the samples that resulted in the best size distribution and best visual appearance.

Experiment	Laps	Oil phase	Span 20	Tween 80	Variable	Deionized water	PSI
Delta control	10	9.86%	1.45%	4.89%	Mesquite gum: 4.93%	78.86%	20,000
Control 2	3	9.86%	1.45%	4.89%	Arabic gum: 4.93%	78.86%	20,000
Control 3	10	9.86%	1.45%	4.89%	Span 20: 1.12% + Tween 80: 3.81%	8.86%	20,000
Control 4	5	9.86%	1.45%	4.89%	Deionized water	83.76%	20,000

Table 2.
Details of the formulation of nano-emulsions.

4. Characterization of nano-emulsions

To demonstrate which formulation leads to an optimization of the use of the microfluidizer and natural gums in the formation of these nano-emulsions, it is necessary to evaluate each system previously described in **Table 2**.

4.1 Dynamic light scattering

When comparing these controls against the Delta Control nano-emulsion in **Figure 8** it is observed that Control 3 with 10 turns in the homogenizer presents a smaller size, in this case a single population with a size of 19 nm, but it has a higher polydispersity index (PDI of 0.143). This is attributed to its composition based solely on surfactants Tween 80 and Span 20 with high HLB. On the other hand, Control 4 nano-emulsion, which does not contain gum, and only carries 5% of surfactants Tween 80/Span 20, had a droplet size of 25.3 nm, a lower size than Control Delta but larger than Control 3. With these experiments, it can be implied that the components responsible for the small drop size are the surfactants Tween 80 and Span 20, since greater interfacial activity with the presence of these surfactants was expected. In addition to increasing the concentration of these surfactants (Control 3) produces a greater interfacial area and a smaller drop size. Finally, it is observed that the Control 2 experiment (control using Arabic gum instead of mesquite gum), has a droplet size of 46.8 nm, very similar to Control Delta, which shows that the mesquite gum, in effect, has a very similar performance to that of Arabic gum (in terms of the droplet size). It should be noted that Control Delta nano-emulsion (with mesquite gum) has a narrower size distribution than Control 2 (with Arabic gum).

A small initial droplet size is not a guarantee that the kinetic stability will be better compared to nano-emulsions with a larger droplet size. For this reason, nano-emulsions were monitored in order to see if there was an increase in the droplet size or an increase in the number of populations, which would indicate instability or some other problem such as cremation, sedimentation, flocculation, or some change in coloration or appearance, which would reduce shell life. All the samples were refrigerated at 4°C and were wrapped in aluminum foil in order to prevent the citrus essential oil from oxidizing in the presence of light.

The first monitoring study to be discussed is Control Delta (**Figure 9**), which represents the best formulation that includes mesquite gum. It was observed to be stable for 4 months. The PDI varied between 0.071 and 0.091 and the sizes vary from 41 to 46 nm, therefore it is considered to be a very stable nano-emulsion; besides, its appearance including color did not change.

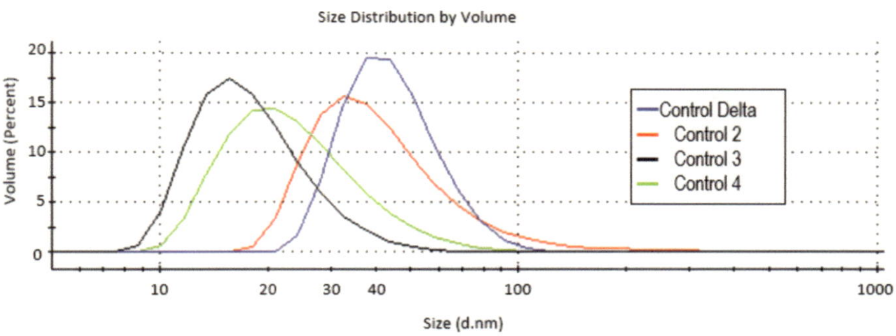

Figure 8.
Graph of volume size distribution in DLS of nano-emulsions.

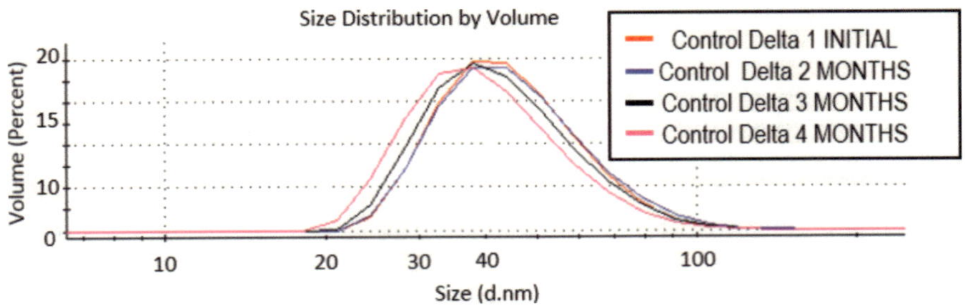

Figure 9.
DLS graph of the 4-month follow-up of Delta Control.

Therefore, a certain percentage of the droplets increased slightly in size, however, there appeared to be no coalescence since the size remained almost constant. On the other hand, the fact that no additional larger populations were produced, unlike the other controls, as shown below, is indicative of the appropriate steric stabilization conferred by mesquite gum.

The next control that was evaluated is the one that incorporates Arabic gum instead of mesquite gum (control 2). From the first series evaluated on this research, we have seen that the nano-emulsions that incorporate Arabic gum developed during this research do not present a very good stability performance, this is confirmed with the results shown in **Figure 10**, where there appears to be an apparent gradual reduction of the droplet size (~25 nm) but with an increase of the PDI from 0.115 to 0.247, to finally increase again at the fourth month (droplet size of ~35 nm). The greatest sign of instability is the presence of other populations with a size around 3 μm that occurs in parallel to the apparent reduction in droplet by the second month, which is attributed to the phenomenon of Ostwald ripening [36], which is one of the main mechanisms of instability in nano-emulsions. This experiment confirms that mesquite gum has advantages over Arabic gum in the formulation of the nano-emulsions of Persian lemon oil. The mesquite gum confers a better steric stabilization as compared to Arabic gum, improving the kinetic stability of the nano-emulsion.

In the follow-up of Control 3 shown in **Figure 11** (nano-emulsion with excess of Tween 80 and Span 20, and without mesquite gum), it seems to be a very stable nano-emulsion, however with each measurement the droplet size, the presence of larger size populations and the PDI increase. For example, from the second month, it evolves from having a single population of 19.0 nm, to having the population of 20.7 nm in coexistence with populations of 467 and 5033 nm, which do not occur in Control Delta. Therefore, it is deduced that the presence of the mesquite gum helps to maintain the stability of the nano-emulsions, providing an additional steric stabilization against coalescence.

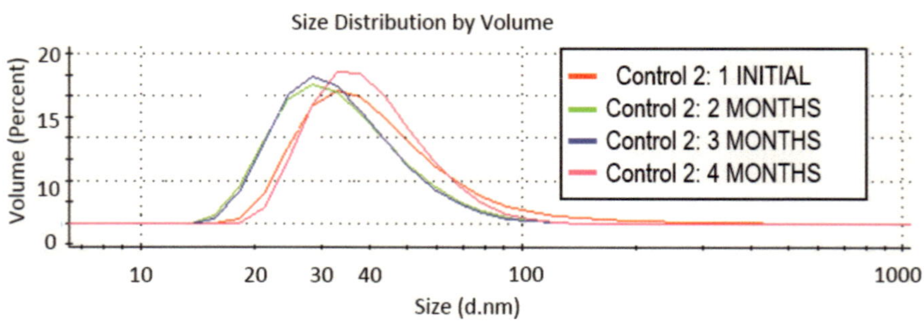

Figure 10.
DLS graph of the 4-month follow-up of Control 2.

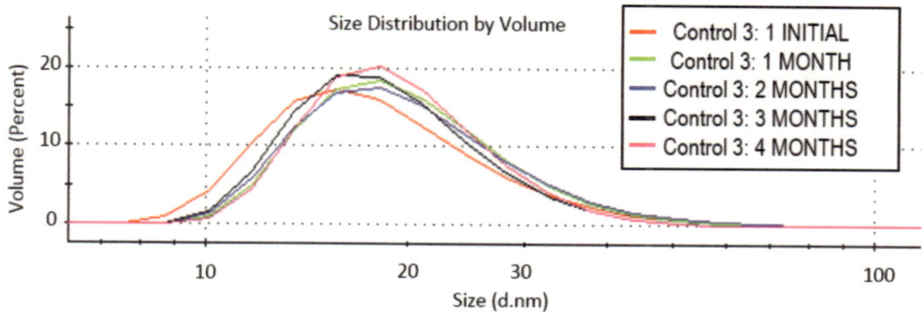

Figure 11.
DLS graph of the 4-month follow-up of Control 3.

Control 4 nano-emulsion incorporates only 5% of surfactant, a mixture of Tween 80/Span 20, and mesquite gum was not included, therefore its water content increases to 85% (as mesquite gum is replaced by water). This control was formulated in order to verify if there is any effect on the size of the nano-emulsion and its stability with the presence of the mesquite gum. This is observed in **Figure 12**, where first, apparent reductions followed by increases in the size of the droplets are seen (droplet size range from ~15 to ~20 nm with PDI range from 0.165 to 0.453), with the presence of large droplet populations, with size around 2000–4000 nm by the fourth month. This may be attributed to a combination of Ostwald ripening and coalescence destabilization phenomena.

In general, the results from the experiments described above showed that nano-emulsions without mesquite gum result in droplet sizes which are smaller than those obtained with samples that include mesquite gum in their formulation, but better kinetic stability and smaller PDI are obtained with the nano-emulsions that contain mesquite gum. This is attributed to the additional steric stability conferred by mesquite gum, which results it a better kinetic stability, since there is virtually no droplet size growth. An arrangement of the different surfactants and stabilizers at the droplet interface is proposed in the scheme of **Figure 13**. It is proposed that the larger hydrodynamic size obtained with mesquite gum is due to its location at the surface of the droplet, where carbohydrate chains of the gum can interact with the sorbitan groups and EO groups of Tween 80.

4.2 Minimum inhibitory concentration

Next, the results of the antibacterial activity tests of Control Delta, Control 2, Control 3 and Control 4 nano-emulsions are shown (in this study, each

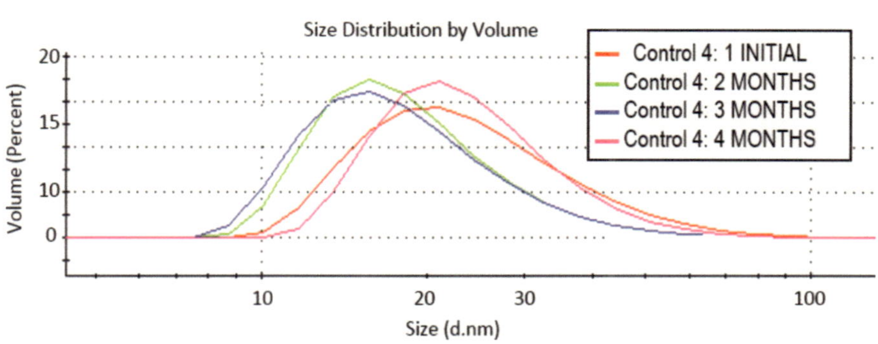

Figure 12.
DLS graph of the 4-month follow-up of Control 4.

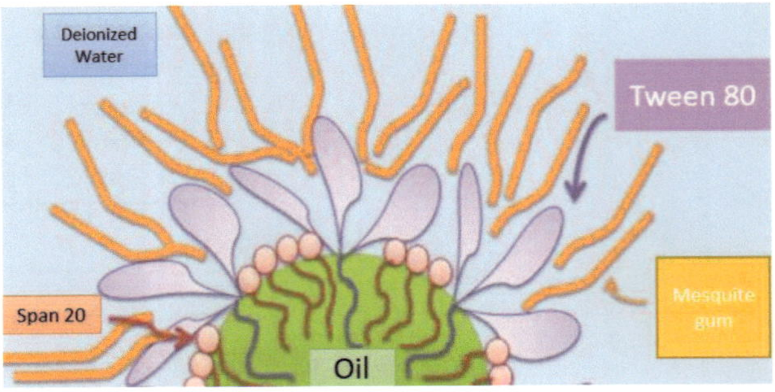

Figure 13.
Schematic of the proposed arrangement of surfactants and stabilizers at the interface of the nano-emulsion droplets.

nano-emulsion was evaluated at 1, 3, 5 and 10 steps into the microfluidizer). To determine the MIC (minimum inhibitory concentration) of the nano-emulsions against the test organisms *Escherichia coli* and *Staphylococcus aureus* the broth micro-dilution method was used as recommended by the National Committee for Clinical Laboratory standards. This test was performed in sterile 96-well microplates. The nano-emulsions were properly prepared and transferred to each microplate into two lines in order to verify reproducibility. The inoculate (10 μL) containing 5x105 CFU (colony-forming unit) of each microorganism was added to each well. A number of wells were reserved in each plate to test for sterility control (no inoculate added), inoculate viability (no nano-emulsion added), and the nano-emulsion inhibitory effect. Plates were aerobically incubated at 35°C. After incubation for 18–24 h, bacterial growth was evaluated by the presence of turbidity and a pellet formed at the bottom of the well. MIC was defined as the lowest concentration of nano-emulsions that had no macroscopically visible growth. A sterilization process was applied to the nano-emulsion samples prior to MIC studies, in order to ensure that no previous contamination was present in the samples.

Table 3 shows the MIC results for nano-emulsions sterilized during 40 minutes under UV light corresponding to *Escherichia coli* and **Table 4** shows the MIC results corresponding to *Staphylococcus aureus*. Additionally, a nano-emulsion with the same composition and processing of Control Delta was prepared, but replacing the Persian lemon essential oil with industrial D-limonene, since this component could be the active bactericidal component.

In general, the best result of MIC was obtained with *Staphylococcus aureus*. The delta control nano-emulsion resulted in a MIC of 6.25% for both bacteria. Therefore, with these results it was confirmed that the nano-emulsions of Persian lemon oil developed under the method described in this research have an antibacterial effect against *Staphylococcus aureus* and *Escherichia coli*.

Nano-emulsion	Steps in Homogenizer	MIC (% of concentration of the nano-emulsion)
Control Limonene	All steps	25
Control Delta	All steps	6.25
Control 2	All steps	25
Control 3	1, 3, 5 10	6.25 12.5
Control 4	All steps	6.25

Table 3.
MIC results for nano-emulsions sterilized during 40 minutes. Bacterium used: ***Escherichia coli.***

Nano-emulsion	Steps in Homogenizer	MIC (% of concentration of the nano-emulsion)
Control Limonene	All steps	25
Control Delta	All steps	6.25
Control 2	All steps	25
Control 3	1 3 5 10	1.56 3.12 6.25 12.5
Control 4	1, 3 5, 10	3.12 6.25

Table 4.
MIC results for nano-emulsions sterilized during 40 minutes. Bacteria used: ***Staphylococcus aureus.***

Considering that the nano-emulsions contain 10% Persian lemon oil, the MIC of this essential oil could be considered as 0.625% for both *Staphylococcus aureus* and *Escherichia coli* and (taking into account the composition of Control Delta samples). Considering the best results, we have a MIC of 0.625% for Control Delta, Control 3 and Control 4 for *Escherichia coli* and 0.156% for Control 3 and *Staphylococcus aureus*; it is inferred then, that these nano-emulsions (Control 2, 3 and 4), which present smaller droplet sizes, their antibacterial power can be attributed to a greater interfacial area, since there is a greater contact area between Persian lemon oil and bacteria. However, due to the better kinetic stability of Control Delta nano-emulsion, it is considered as more promising. Some adjustments could be made to improve these results, such as increasing the concentration of Tween 80/Span 20 in Control Delta in order to reduce the size, but maintaining the presence of mesquite gum in order to preserve the steric stability conferred by it.

When carrying out the antibacterial activity tests, it was observed that when the nano-emulsions were subjected to treatment with UV light, they became slightly more transparent, so it was suspected that the UV light radiation can cause a reduction in the droplet size. In **Figures 14** and **15**, we observe the effect of the UV light treatment on the two nano-emulsions with better kinetic stability behavior and droplet size. In the case of Control 3 nano-emulsion at 10 steps (formulation without mesquite gum, but with additional Tween 80 and Span 20), there was no significant reduction in size (from 1 to 0.5 nm reduction), after sterilization treatment at different exposure times. However, for Control Delta nano-emulsion there was a reduction of approximately 10 nm after treatment with UV light, which may indicate that the mesquite gum is being broken into smaller carbohydrate chains, or

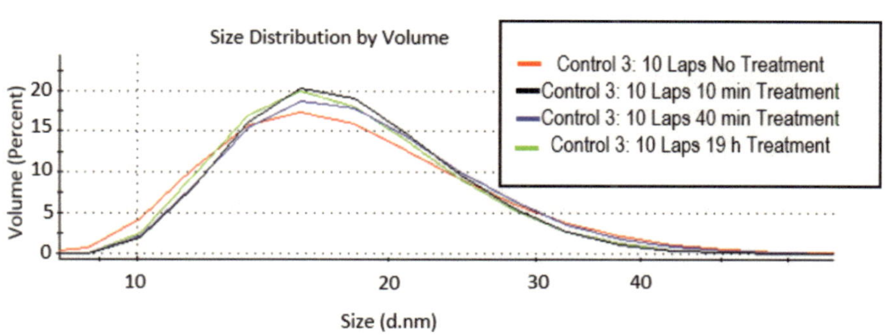

Figure 14.
Size distribution chart by volume of control 3 10 steps with UV treatment.

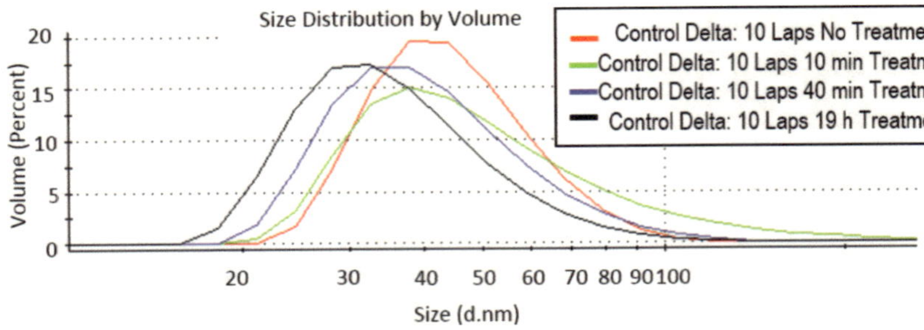

Figure 15.
Size distribution graph by volume of control delta 10 steps with UV treatment.

that rearrangement of the carbohydrate chains is taking place, thereby reducing the hydrodynamic droplet size.

On the other hand, although the literature indicates that it is not that clear which of the components of the citrus essential oils is the cause of the antibacterial effect, the nano-emulsion with industrial D-limonene results in a higher MIC than Control Delta which is prepared with Persian lemon oil; thus, it may be inferred that the aldehydes of the Persian lemon oil could be the components that are mainly responsible for this effect, as compared to the terpene components.

4.3 Volatile profile by gas chromatography of nano-emulsions

After the analysis of the previous results was carried out, it may be inferred that the aldehydes present in the Persian lemon oil may be the components that potentiate the antibacterial power. Nano-emulsions of Control Delta (Combination of Surfactants—Span 20: 1.45%, Tween 80: 4.9%, Mesquite Gum 4.9%), both unsterile and sterilized (with 10 and 40 minutes of UV light treatment) were investigated in order to perform a profile analysis of the volatiles by gas chromatography. As observed in **Figure 16**, when the nano-emulsion Control Delta was exposed to different UV sterilization times, the D-limonene signal becomes smaller in comparison with the nano-emulsion sample that was not exposed to UV, so it can be implied that the D-limonene component does not appear to be the main component that acts as an antibacterial agent in this particular sample.

Once the retention time of D-limonene passes (**Figure 17**), the region of the aldehydes begins at higher retention times, and as observed in the previous analysis

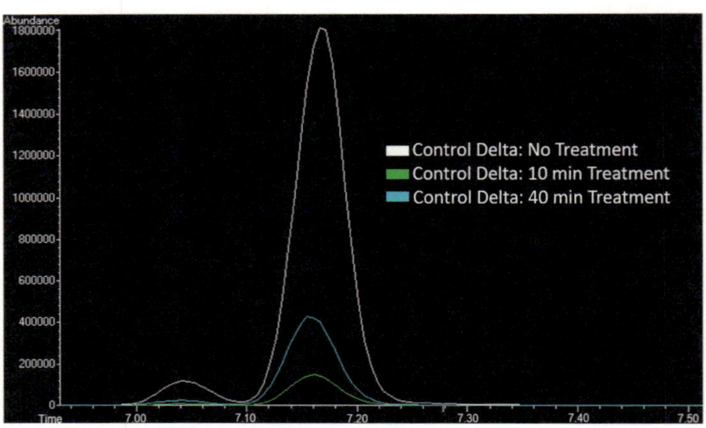

Figure 16.
D-limonene signal in the Control Delta Chromatogram.

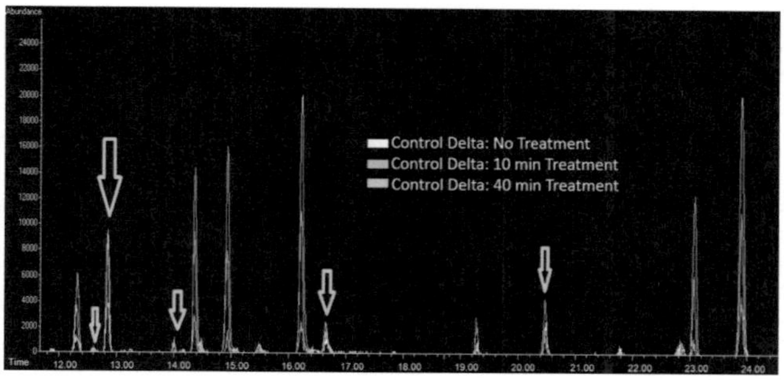

Figure 17.
Central area of the Control Delta Chromatogram.

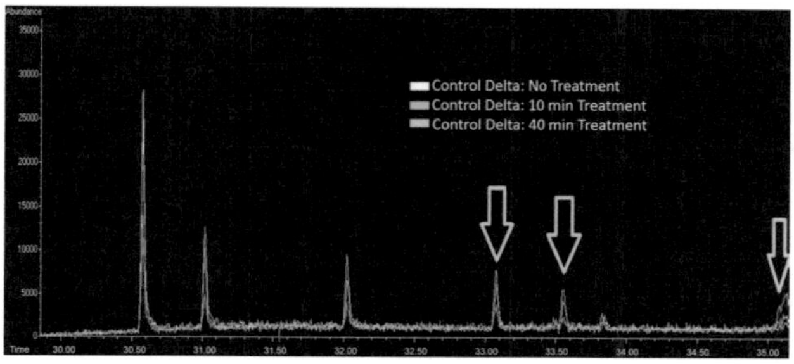

Figure 18.
Oxides area in the Delta Control Chromatogram.

of the Persian lemon essential oil, it contains a higher concentration of aldehydes than other citrus essentials oils. There are several aldehyde signals that increase when the sample is subjected to 40 minutes of UV. It was observed that the nano-emulsions contain a higher percentage of aldehydes and a better antibacterial response after UV treatment.

The oxides zone can be detected after a retention time of 30 minutes (**Figure 18**). In this area, other signals increase when the nano-emulsions are subjected to 40 minutes sterilization process. Thus, it may be inferred that these oxides may also contribute to the antibacterial effect.

5. Conclusions

This chapter examines the influence of nano-emulsion composition and high pressure homogenization conditions on droplet size and stability. Nano-emulsions with a droplet size smaller than 100 nm (diameter) can be produced by precise conditions of pressure, a specific number of steps in the high pressure homogenizer, and the presence of a combination of surfactants and emulsifiers that is capable to perform ideally under those conditions without degrading. Surfactants and natural gums were used to produce nano-emulsions with small droplets of Persian lemon oil, which has antimicrobial properties, with potential applications in the areas of cosmetics, pharmaceuticals and the food industry. In this particular study mesquite gum was used for the first time to maintain the kinetic stability of the Persian lemon oil droplets.

With the results of the MIC characterization it was confirmed that the nano-emulsions of Persian lemon oil developed under the method described in this research have an antibacterial effect against *Staphylococcus aureus* and *Escherichia coli*. And although there is controversy about which components of the citric essential oils increase the antibacterial activity, our study suggest that aldehydes have an important role for the antibacterial effect of Persian lemon oil nano-emulsions.

Considering the series of experiments described in this research and the study of stability as a function of time it is clear that the Control Delta formulation composed of Span 20 (1.45%), Tween 80 (4.9%) and mesquite gum (4.9%) was the best. It was also shown that mesquite gum is superior to Arabic gum for the kinetic stability as shown by the behavior of the nano-emulsion that replaces mesquite gum with Arabic gum (Control 2). Indeed, when an excess of surfactant (Tween 80 and Span 20) is used, it is possible to obtain a smaller droplet size as observed in other control (Control 3 with excess Tween 80 and Span 20), but mesquite gum provides additional steric stabilization that confers a better stability over time against coalescence. The sample that replaces mesquite gum with water (Control 4) confirms the additional steric stability of mesquite gum.

Acknowledgements

M. B. M. T. acknowledges financial support (Master degree grant) from CONACYT. M. B. M. T. acknowledges Lilia Bautista from CIMAV Monterrey for their help with FT-IR and DLS, measurements and assistance, respectively. M. B. M. T acknowledges M.Sc. Lilia Torres from CIATEJ Noroeste for all the assistance. M. B. M. T acknowledges Ph.D. Guillermo Acosta and Q. Ashanti Rodríguez for their help with the analysis of the minimum inhibitory concentration (MIC) of the nanoemulsions. We acknowledge Frutech International Corporation for their donation of the citric essential oils.

Conflict of interest

The authors declare no conflict of interest.

Author details

Maira Berenice Moreno-Trejo[1], Arturo Adrián Rodríguez-Rodríguez[1], Ángela Suarez-Jacobo[2] and Margarita Sánchez-Domínguez[1]*

1 Centro de Investigación en Materiales Avanzados S.C. (CIMAV) Unidad Monterrey, Apodaca, Nuevo León, Mexico

2 Centro de Investigación y Asistencia en Tecnología y Diseño del Estado de Jalisco, A.C. Unidad Noreste, Apodaca, Nuevo León, Mexico

*Address all correspondence to: margarita.sanchez@cimav.edu.mx

References

[1] McClements DJ. Nanoemulsions versus microemulsions: Terminology, differences, and similarities. Soft Matter. 2012;**8**(6):1719. Available from: http://pubs.rsc.org/en/content/articlehtml/2012/sm/c2sm06903b

[2] Izquierdo P, Esquena J, Tadros TF, Dederen C, Garcia MJ, Azemar N, et al. Formation and stability of nano-emulsions prepared using the phase inversion temperature method. Langmuir. 2002;**18**(1):26-30

[3] Solans C, Izquierdo P, Nolla J, Azemar N, Garcia-Celma MJ. Nano-emulsions. Current Opinion in Colloid & Interface Science. 2005;**10**(3-4):102-110

[4] Forgiarini A, Esquena J, González C, Solans C. Studies of the relation between phase behavior and emulsification methods with nanoemulsion formation. In: Trends in Colloid and Interface Science XIV. Berlin, Heidelberg: Springer; 2000. pp. 36-39

[5] Becher P. Encyclopedia of Emulsion Technology: Vol. 1. Basic Theory. New York: Marcel Dekker; 1983

[6] Tadros TF. Applied Surfactants: Principles and Applications. Weinheim: Wiley-VCH Verlag GmbH & Co. KGaA; 2005

[7] Sharma N, Bansal M, Visht S, Sharma PK, Kulkarni GT. Nanoemulsion: A new concept of delivery system. Chronicles of Young Scientists. 2010;**1**(2):2-6

[8] Shinoda K. The correlation between the dissolution state of nonionic surfactant and the type of dispersion stabilized with the surfactant. Journal of Colloid and Interface Science. 1967;**24**(1):4-9

[9] Shinoda K, Saito H. The effect of temperature on the phase equilibria and the types of dispersions of the ternary system composed of water, cyclohexane, and nonionic surfactant. Journal of Colloid and Interface Science. 1968;**26**(1):70-74

[10] Sarker DK. Engineering of nanoemulsions for drug delivery. Current Drug Delivery. 2005;**2**(4):297-310

[11] Florence AT, Whitehill D. Stabilization of water/oil/water multiple emulsions by polymerization of the aqueous phases. The Journal of Pharmacy and Pharmacology. 1982;**34**(11):687-691

[12] Moreno-Trejo MB, Sánchez-Domínguez M. Mesquite gum as a novel reducing and stabilizing agent for modified tollens synthesis of highly concentrated Ag nanoparticles. Materials. 2016;**9**(10):817

[13] Sanchez-Dominguez M, Pemartin K, Boutonnet M. Preparation of inorganic nanoparticles in oil-in-water microemulsions: A soft and versatile approach. Current Opinion in Colloid & Interface Science. 2012;**17**(5):297-305. DOI: 10.1016/j.cocis.2012.06.007

[14] Salager JL, Loaiza-Maldonado I, Minana-Perez M, Silva F. Surfactant-oil-water systems near the affinity inversion part I: Relationship between equilibrium phase behavior and emulsion type and stability. Journal of Dispersion Science and Technology. 1982;**3**(3):279-292

[15] White EV. The constitution of mesquite gum. I. The methanolysis products of methylated mesquite gum. Journal of the American Chemical Society. 1946;**68**(2):272-275

[16] Aspinall GO, Whitehead CC. Mesquite gum. I. The

4-O-methylglucuronogalactan core. Canadian Journal of Chemistry. 1970;**48**(24):3840-3849

[17] López-Franco YL, Córdova-Moreno RE, Goycoolea FM, Valdez M a, Juárez-Onofre J, Lizardi-Mendoza J. Classification and physicochemical characterization of mesquite gum (Prosopis spp.). Food Hydrocolloids. 2012;**26**(1):159-166. Available from: http://linkinghub.elsevier.com/retrieve/pii/S0268005X11001615

[18] Osman ME, Williams PA, Menzies AR, Phillips GO. Characterization of commercial samples of gum arabic. Journal of Agricultural and Food Chemistry. 1993;**41**(1):71-77

[19] Viuda-Martos M, Ruiz-Navajas Y, Fernández-López J, Pérez-Álvarez J. Antifungal activity of lemon (*Citrus lemon* L.), mandarin (*Citrus reticulata* L.), grapefruit (*Citrus paradisi* L.) and orange (*Citrus sinensis* L.) essential oils. Food Control. 2008;**19**(12):1130-1138

[20] Lee H, Cheng S, Chang S. Antifungal property of the essential oils and their constituents from Cinnamomum osmophloeum leaf against tree pathogenic fungi. Journal of the Science of Food and Agriculture. 2005;**85**(12):2047-2053

[21] Garcia R, Alves ESS, Santos MP, Aquije GMF, Fernandes AAR, dos SRB, et al. Antimicrobial activity and potential use of monoterpenes as tropical fruits preservatives. Brazilian Journal of Microbiology. 2008;**39**(1):163-168

[22] Uysal B, Sozmen F, Aktas O, Oksal BS, Kose EO. Essential oil composition and antibacterial activity of the grapefruit (*Citrus paradisi*. L) peel essential oils obtained by solvent-free microwave extraction: Comparison with hydrodistillation. International Journal of Food Science and Technology. 2011;**46**(7):1455-1461

[23] Sanei-Dehkordi A, Sedaghat MM, Vatandoost H, Abai MR. Chemical compositions of the peel essential oil of Citrus aurantium and its natural larvicidal activity against the malaria vector Anopheles stephensi (Diptera: Culicidae) in comparison with Citrus paradisi. Journal of Arthropod-Borne Diseases. 2016;**10**(4):577

[24] Schlüter B, Pflegel P, Lindequist U, Jülich WD. Aspects of the antimicrobial efficacy of grapefruit seed extract and its relation to preservative substances contained. Die Pharmazie. 1999;**54**(6):452-456

[25] Sierra PO. Manual sobre El Cultivo de Limón Persa (citrus latifolia tanaka) en Guatemala. Man sobre El Cultiv Limón Persa (citrus latifolia tanaka) en Guatemala. Inst Cienc y Tecnol Agrícola–IICTA-Guatemala. 2002;12

[26] Li P-H, Chiang B-H. Process optimization and stability of D-limonene-in-water nanoemulsions prepared by ultrasonic emulsification using response surface methodology. Ultrasonics Sonochemistry. 2012;**19**(1):192-197

[27] Donsì F, Annunziata M, Vincensi M, Ferrari G. Design of nanoemulsion-based delivery systems of natural antimicrobials: Effect of the emulsifier. Journal of Biotechnology. 2012;**159**(4):342-350

[28] Donsì F, Annunziata M, Sessa M, Ferrari G. Nanoencapsulation of essential oils to enhance their antimicrobial activity in foods. LWT- Food Science and Technology. 2011;**44**(9):1908-1914

[29] Gamarra FMC, Sakanaka LS, Tambourgi EB, Cabral FA. Influence on the quality of essential lemon (*Citrus aurantifolia*) oil by distillation process. Brazilian Journal of Chemical Engineering. 2006;**23**(1):147-151

[30] Hernández-Ochoa L, Gonzales-Gonzales A, Gutiérrez-Mendez N, Muñoz-Castellanos LN, Quintero-Ramos A. Estudio de la actividad antibacteriana de películas elaboradas con quitosano a diferentes pesos moleculares incorporando aceites esenciales y extractos de especias como agentes antimicrobianos. Revista Mexicana de Ingenieria Quimica. 2011;**10**(3):455-463

[31] Ramos-García M d L, Bautista-Baños S, Barrera-Necha LL, Bosquez-Molina E, Alia-Tejacal I, Estrada-Carrillo M. Compuestos antimicrobianos adicionados en recubrimientos comestibles para uso en productos hortofrutícolas. Revista Mexicana de Fitopatología. 2010;**28**(1):44-57

[32] Montes-Belmont R. Diversidad de compuestos químicos producidos por las plantas contra hongos fitopatógenos. Revista Mexicana de Micología. 2009;**29**:73-82

[33] Nannapaneni R, Muthaiyan A, Crandall PG, Johnson MG, O'Bryan CA, Chalova VI, et al. Antimicrobial activity of commercial citrus-based natural extracts against Escherichia coli O157: H7 isolates and mutant strains. Foodborne Pathogens and Disease. 2008;**5**(5):695-699

[34] Bakkali F, Averbeck S, Averbeck D, Idaomar M. Biological effects of essential oils–a review. Food and Chemical Toxicology. 2008;**46**(2):446-475

[35] Kalemba D, Kunicka A. Antibacterial and antifungal properties of essential oils. Current Medicinal Chemistry. 2003;**10**(10):813-829

[36] Wooster TJ, Golding M, Sanguansri P. Impact of oil type on nanoemulsion formation and Ostwald ripening stability. Langmuir. 2008;**24**(22):12758-12765

Chapter 5

An Update on Nanoemulsions Using Nanosized Liquid in Liquid Colloidal Systems

Praveen Kumar Gupta, Nividha Bhandari, Hardik N. Shah, Vartika Khanchandani, R. Keerthana, Vidhyavathy Nagarajan and Lingayya Hiremath

Abstract

Nanoemulsions, kinetically stable and thermodynamically unstable colloidal liquid-in-liquid dispersions with droplet sizes in the order of 20–500 nm mainly consist of oil, surfactants, co surfactants and an aqueous phase. There are various methods for the fabrication of Nano-emulsions which can be divided based on the energy required—High energy emulsification methods and Low energy emulsification methods. High energy emulsification includes methods like Ultra sonication, high pressure homogenization using either microfluidizers or high-pressure homogenizers. Low energy emulsification has drawn attention since they are soft, nondestructive and cause no damage to encapsulated molecules and includes methods like phase inversion temperature, solvent displacement, phase inversion composition method. Nanoemulsions are best suited for drug delivery systems because of their lipophilic nature, optical clarity and surface area. Owing to their nature to prevent flocculation and inherent creaming, nanoemulsions find an important place in the cosmetic industry also. This chapter provides an insight into the use of nanogels, emulsion based wet wipes and PEG free nanoemulsions in cosmetics. In the food industry, nanoemulsions are utilized for the production of functional foods. Some of the patented nanoemulsions and their commercial applications have also been mentioned.

Keywords: nanoemulsions, high pressure homogenization, ultrasonication, phase inversion temperature, solvent displacement, phase inversion composition, drug delivery, penetration pathways, cosmetics, nanogels, PEG free emulsions, wet wipes, functional foods, patented nanoemulsions

1. Introduction

In this chapter we will be briefing about the emulsion and its types. We will also be discussing in detail about nanoemulsions, its types, fabrication, application and its properties.

1.1 Emulsion systems and its types

An emulsion system generally consists of two or more liquids that are immiscible. They are a type of colloids, which are two-phase systems of matter. In emulsion systems, the two phases are dispersed and continuous. One liquid is dispersed (the disperse phase) in the other (the continuous). The different types of emulsion systems can include: oil-in-water (o/w), water-in-oil (w/o), and oil-in-oil (o/o) [1]. The oil-in-oil phase can be polar oil dispersed in non-polar oil, or vice versa. An emulsifier is usually used to disperse immiscible liquids. The emulsifier also plays an important role in the formation and long-term stability of the emulsions. Emulsions can also be classified on the type of emulsifier or the structure of the system. Emulsions being liquids do not have any static internal structure. The droplets are assumed to be statistically distributed in the liquid matrix. According to IUPAC, in emulsions the droplets can be amorphous, liquid-crystalline, or any mixture. The droplet diameters in the dispersed phase range between 10 nm and 100 μm (which may exceed the size limits for colloidal particles) [2]. Some common example of emulsion systems are homogenized milk, some cutting fluids for metal working, egg yolk is an emulsion with the emulsifying agent lecithin, butter is an emulsion of water in fat, and an emulsion of silver halide in gelatin is used as a coating in the photosensitive side of a photographic film [3].

Emulsion system can be classified based on their droplet size as macroemulsion, nanoemulsion, and miniemulsion (**Table 1**).

Emulsion systems find a wide range of applications in the field of food, cosmetics, agriculture, pharmaceuticals (preparation of drugs and drug delivery).

1.2 Nanoemulsions

As the name suggests, the size of the droplets in this type of emulsion is in nanometer ranges. They not only differ in size but also in the many properties and method of preparation. The main difference between nanoemulsion and conventional emulsion (macroemulsion) is the size and shape of the droplets in the continuous phase. In macroemulsion, the shape is usually spherical but in nanoemulsions a variety of shapes can be seen like swollen micelles and bicontinuous structures. Though micro and nanoemulsions are similar in their sizes the method of preparation differs between them. Both of them require energy inputs, but nanoemulsions mostly use mechanical shear while micro emulsions make use of spontaneous emulsification methods. Microemulsions also need a high surfactant concentration compared to nanoemulsions. The application of nanoemulsion in pharmaceutical, food, cosmetic, and chemical industry is comparatively more than microemulsion since moderate surfactant concentration is sufficient for their making [5].

Nanoemulsions are said to be kinetically stable and thermodynamically unstable. Their stability can be altered by their preparation methods like adding specific co-surfactants. They usually use high energy methods for their preparation but low energy based methods can also be used with the help of some special conditions using certain chemical potential of the component [6]. Nanoemulsions are said to be transparent, biodegradable, and biocompatible. Normal emulsions usually undergo sedimentation by gravity, which is overcome by nanoemulsions. Nanoemulsions exhibit Ostwald ripening phenomenon. Due to this, small

Emulsion type	Droplet size (μm)
Macroemulsion	1–100
Microemulsion	10–100
Nanoemulsion	20–500

Table 1.
Emulsion type and its droplet size [4].

molecules collide and form large globules. To overcome this, co-surfactants are added or second oil is added to the dispersion phase. Proper manufacturing procedure also helps overcome Ostwald ripening [7]. Nanoemulsions provide a wide surface area and so allow active components to penetrate easily and faster. Another important characteristic of nanoemulsion is their transparent optical property. This is mainly due to their size, which is one fourth of the wavelength of visible light [8]. Nanoemulsions have the ability to solubilize both hydrophobic and hydrophilic substances, and hence enhance their permeability and bioavailability [9]. This makes them very useful as drug delivery systems for both the type of drugs.

Nanoemulsions are also said to have tunable rheological properties. They are tuned by controlling the dispersed phase volume, droplet size or the addition of salt and depletion agents [4]. Hence nanoemulsions can be tuned from being a free flowing fluid to a gel like substance [10]. Addition of polymers also tunes the rheological properties. The polymers associate either with themselves or with the nanoemulsions. A thermo reversible gel was made, where a polymer gelator (with two hydrophobic end groups) was added. At temperatures greater than the gelling temperature, the polymer's two hydrophobic ends bridges with the nanoemulsion droplets making them a gel. At lower temperature, they detach and hence return to a transparent fluid like structure.

There are three types of nanoemulsion based on the composition:

- Oil in water: oil droplets are dispersed in continuous aqueous phase.
- Water in oil: water droplets are dispersed in continuous oil phase.
- Bi-continuous (double): micro domains of oil and water are interdispersed within the system [11].

1.3 Current trends in nanoemulsions

As already mentioned, nanoemulsions are being used in a wide range of fields. There is a lot of research and development work done in the field of nanoemulsions. Many bioactive substances are present in natural available substances, emulsification of these bioactive components is a trending research topic. In September 2018, water compatible form of coconut oil through nano-emulsification was developed [12]. The nanoemulsion was made successfully using PHC as a surfactant at a concentration of 5% {w/w}. Nanoemulsions have also found an important space in field of pharmaceuticals. Many of the oral drugs synthesized do not have aqueous stability almost insoluble) and have low bioavailability. A low energy method to make composite hydrogel beads encapsulated with single and multiple hydrophobic drugs was developed [13]. This makes nanoemulsions a promising carrier of hydrophobic drugs. It was shown that nanoemulsions were used to enhance the antileishmanial activity of *Copaifera* spp. oleoresins against both *Leishmania amazonensis* and *Leishmania infantum* strains [14].

Recently a new technique for making Pickering nanoemulsions using Silica nanoparticles was developed which is highly scalable and energy efficient. Nanoemulsions are usually stabilized using surfactants. The use of surfactants has some disadvantages which include surfactant desorption and Ostwald ripening. Hence a new interest of making nanoparticle stabilized nanoemulsion (Pickering nanoemulsions) has evolved. Nanoparticles have higher desorption energy barrier. However, the limitation of nanoparticles as stabilizing agent was obtaining the size in nano-range. In the traditional method to make Pickering emulsions (high energy and low energy methods, described in the next section) many steps were involved, a single step method was developed using vapor condensation. Moreover the traditional methods used in the preparation of Pickering nanoemulsions had some disadvantages. High energy methods reduced the adsorption of the particles on the droplets while, low energy methods were unable to produce Pickering nanoemulsions and clogging of nanoparticles was seen. The concentration of nanoparticles required in the new methods was also less compared to that required in traditional methods. In this process, oil was taken and cooled below the dew point, during which the water condenses on the oil. If the oil has the right properties and sufficient concentration of nanoparticles, then water drops self-disperse within the oil. The nanoparticles then will self-assemble around them to form nanoemulsions [15]. Nanoemulsions are studied in great detail due to their potential applications. Improvements in their preparation methods and the fields in which they can be used are the ongoing trends in nanoemulsions.

2. Fabrication

The fabrication of nanoemulsion involves the preparation of macroemulsions and then it's conversion to nanoemulsion by various methods, all of which can be categorized into either Low energy or High energy methods [5, 16]. Techniques which involve modification of factors responsible for the hydrophilic–lipophilic balance come under Low energy methods and those that use mechanical devices to break down the particles to small sizes are referred to as high energy methods. As much as composition is responsible for the properties of the nanoemulsion so is the technique used for its preparation. In this section a brief insight is given on a few widely used methodologies.

2.1 High energy methods for emulsion formation

In contrast to the low energy methods for nanoemulsion formation, high energy methods require the use of many devices which uses mechanical or chemical energy as input to form small droplets because they are non-equilibrium systems which cannot be formed spontaneously [17]. These devices often entail a huge initial cost as well as expenses to maintain throughout use. The purpose of these devices in high energy methods is to provide intense mechanical energy that helps to break up macroscopic phases or turn larger droplets into smaller droplets [18]. These devices provide forces so strong that it disrupt water and oil phases to form nanoemulsions. In high energy methods, input energy density is about 108–1000 W/kg. The required energy supplied is in very shortest duration of time to the system in order to obtain homogeneous small sized particles. In addition to this, the high energy methods for 55 nanoemulsion formation are not limited by the types of oil and emulsifiers that can be used like the low energy methods are. At present high energy methods are more frequently utilized in the food industry than low energy methods with high pressure valve homogenization, microfluidization, and sonication being the most common [19]. All this high energy methods are impacted by emulsion component characteristics (i.e. oil, type, surfactant type, surfactant concentration, viscosity, etc.) and equipment characteristics (i.e. size of the equipment, pressure used, number of passes/time in equipment, design, etc.). The input energy density is about 108–1010 W/kg [20]. These parameters should be optimized for each and every system and high energy method.

2.1.1 High pressure valve homogenizer (HPVH)

HPVH is the most popular method used for the production of nanoemulsions. The most common use is in applications from ketchup processing to milk homogenization and to manufacture nanoemulsions that particle sizes are up to 1 nm [18]. When using a HPVH, a coarse emulsion is initially made using a high-speed mixer, fed into the input valve of the HPVH, and then flowed between the valve seat and valve at a high velocity. The macroemulsion is forced to pass through a small orifice at an operating pressure between 500 and 5000 Psi [18]. Since several forces like hydraulic shear, intense turbulence and cavitation act together extremely small droplet sized nanoemulsions are achieved. The process is repeated till the final product reaches our desired droplet size and polydispersity index (PDI). Lower the PDI means higher uniformity of droplet size in nanoemulsions. Mono-disperse samples have PDI lesser than 0.08, narrow size distribution range is 0.08–0.3 and PDI greater than 0.3 indicates broad size distribution [18].

With an increase in velocity, the pressure decreases causing an instantaneous pressure drop and encouraging the coarse emulsion to impinge on the impact ring [21]. Sometimes HPVH passes through two valves and thus emulsion production will break up into two stages: in the first stage the droplets are broken up and in the second stage a lower pressure is utilized to disrupt any 'flocs' formed by the initial valve [22]. Obtaining submicron levels requires large amount of energy and high temperature which can deteriorate the components. Thermolabile compounds like proteins, enzymes and nucleic acids may be damaged easily [18] (**Figure 1**).

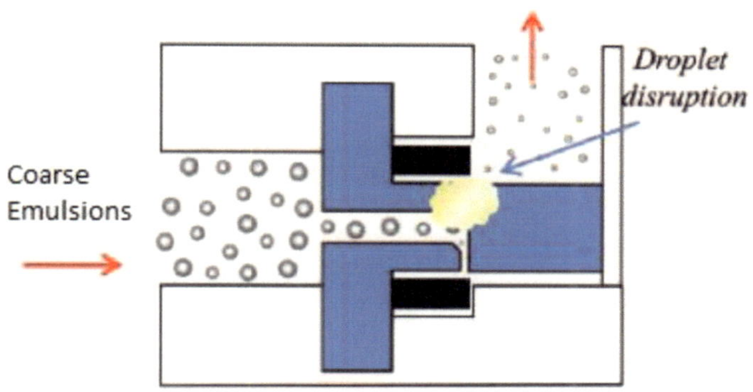

Figure 1.
Schematic representation of high pressure valve homogenizer [18].

2.1.2 Sonication

Emulsions produced by sonication use ultrasonic homogenizers (UH) to provide high intensity of ultrasonic waves to the sample. The frequency of the waves (29 kHz or larger) is higher than the maximum audible frequency of human ear (16–18 kHz) [17]. These waves provide around 56 disruptive forces to breakup oil and water phases thus forming small droplets on the principle of cavitation. Input energy comes from a sonicator probe which can be directly placed in the sample. There are two mechanisms which take part in sonication [17]. Firstly, acoustic field creates interfacial waves which makes oil phase to disperse in the continuous phase as droplets. Secondly, ultrasound provokes acoustic cavitation which provides formation and collapse of microbubbles respectively since there is a pressure fluctuation

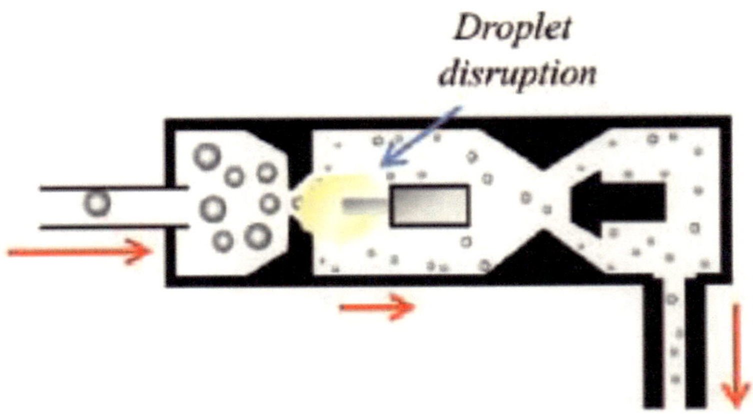

Figure 2.
Schematic representation of ultrasonic jet homogenizer [23].

of a single sound wave [23]. By this enormous levels of highly localized turbulence is generated and it causes micro implosions which disrupt large droplets into sub-micron size. Since most of the ultrasonic systems emits sound field which are inhomogeneous, so in order to have droplets to experience highest shear rate, recirculation of the emulsion through the region of high power must be provided and on repeating recirculation we obtain uniform droplet size at dilute concentration [18].

Presently sonication has been well established for the laboratory scale but it may be difficult to implement on a production scale because of issues like low throughput. Optimization of parameters (like emulsifier type, amount emulsifier and viscosity of phases) is necessary to prepare nanoemulsions having fine droplets [18]. Even, the high local intensity provided by sonication could lead to detrimental quality effects by way of protein denaturation, polysaccharide polymerization or lipid oxidation of the emulsion components [23] (**Figure 2**).

2.1.3 Microfluidization

Microfluidization is the most widely employed and novel technique in the pharmaceutical and cosmetic industry in order to acquire fine emulsions [18]. High pressure is provided by device called Microfluidizers (MF). Initially a coarse emulsion is made using a high speed mixer which is then fed into the hood and accelerated at high velocities within the channels using a pumping device and the macroemulsion to go through the interaction chamber by the high pressure forces and thus nanoemulsions with submicron ranged particles are produced [17]. The channels are made to collide into each other within the interaction chamber [21]. Uniform nanoemulsion can be produced by repeating the process many times and vary the operating pressure to get desired particle size [18].

The main parts of a MF include a fluid inlet (where the coarse emulsion is fed), a pumping device (to help move the emulsion through), and the interaction chamber or nozzle (where the particle collision occurs) [22]. A collision between crude emulsion jets from two opposite channels in the nozzle of microfluidizers is observed. The mobility of crude emulsion is supplied by a pneumatically powered pump that has capability of compressing air up to pressures between 150 and 650 MPa [18]. This high pressure forces the crude emulsion stream to go through microchannels and after the collision of two opposite channels enormous level of shearing force is produced. Hence, by the help of this force fine emulsions are produced [23] (**Figure 3**).

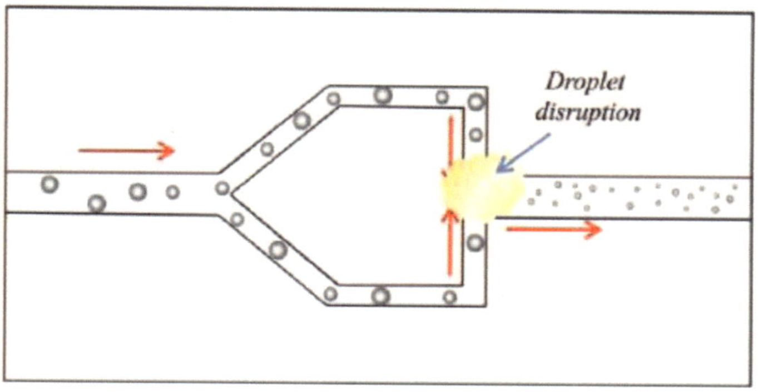

Figure 3.
Schematic representation of microfluidizer [23].

2.2 Low energy methods

Requiring no expensive equipment, easier implementation and better efficiency in terms of energy are the reasons for the growing interest in low energy methods [16, 17, 24, 25]. Moreover, encapsulation of drugs and macromolecules can be carried out due to mild operating temperatures. The necessity for higher amounts of surfactants may be a downside [17]. The whole concept of low energy synthesis has its roots in modification of factors responsible for the hydrophilic–lipophilic balance of the surfactant-oil–water mixture [26]. These include environmental factors like temperature, composition and the chemical potential of the components. Spontaneous Emulsification (SE) and Phase Inversion are two commonly implemented synthesis [17, 24].

2.2.1 Spontaneous emulsification

An emulsion can be fabricated by diluting a biphasic system leading to diffusion of one phase to another. This is usually done by adding the organic phase into the aqueous phase and then a surfactant which is water miscible. The migration of the surfactant causes disorder at the interface of the two phases leading to an increase in the surface area along with the formation of oil droplets in the aqueous phase [16].

To obtain nanoemulsions, the same dilution process is performed on microemulsions. The properties of the nanoemulsion depend on the oil viscosity, surfactant hydrophilic–lipophilic balance and solvent miscibility with water. With the help of an appropriate dilution procedure and composition, both W/O and O/W microemulsions can be used to obtain nanoemulsion. While obtaining it from O/W microemulsion the composition of microemulsion and the procedure of dilution does not matter, whereas while starting with W/O microemulsion the dilution procedure and /or the composition of microemulsion matters. O/W and W/O nanoemulsions can be formed even without a surfactant, this is called the Ouzo effect also known as Solvent displacement method [24, 27, 28]. This phenomenon has mainly been used for fabricating polymeric nanoparticles or nanocapsules using nanoemulsion as a template in drug delivery [25].

2.2.2 Phase inversion

As addressed earlier, there are different types of Nanoemulsions, either oil in water (O/W) or water in oil (W/O). Phase inversion, as the name suggests, is a fabrication method that involves conversion of O/W to W/O emulsion or vice versa. It utilizes the energy released during this conversion for the formation of droplets. This physical process can be brought about by varying the temperature or phase volume ratio, giving rise to phase inversion temperature (PIT) method and phase inversion composition (PIC) method [29].

In these emulsification methods it is very important to know the behavior of the surfactant as it plays a significant role in minimizing both droplet size and polydispersity of the nanoemulsion formed. Its properties also depend on the kinetics of the emulsification process, especially if they have high viscosity [27].

2.2.2.1 Phase inversion temperature

The Phase inversion temperature (PIT) method is used when the surfactants are sensitive to changes in temperature. The principle of this method is based on the changes in surfactant spontaneous curvature (molecular geometry) with temperature. For example, in poly(oxyethylene)-type non-ionic surfactant, increase in temperature causes dehydration of the poly(oxyethylene) chains whereas at low temperature these chains are hydrated and hence are hydrophilic in nature. At one temperature the surfactant exhibits both hydrophilic and lipophilic properties, this temperature is known as HLB temperature (Hydrophilic–Lipophilic Balance) [17, 27]. So, at this temperature the surfactant is equally soluble in the oil and aqueous phase [16, 17].

Using PIT very small sizes of droplets can be obtained. At the HLB temperature due to the low interfacial tension the surfactant forms a layer but as soon as the temperature is changed by quick cooling or heating, the surfactant molecules move from one phase into another resulting in the formation of small oil droplets. The movement of the surfactant molecules depends on its hydrophilicity or lipophilicity of its chain which in turn depends on temperature [16, 17] (**Figure 4**).

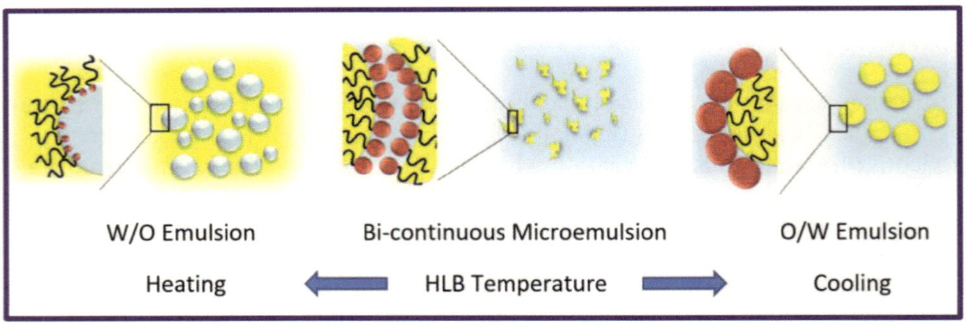

Figure 4.
Shows the curvature of surfactant and the favorable emulsion formed by heating and cooling [16].

2.2.2.2 Phase inversion composition

Phase inversion composition is performed when the surfactant properties changes due to dilution of one of the phases. It involves dilution of oil phase with water or vice versa which causes an increase or decrease in hydration degree of surfactant. In phase inversion composition, phase transition takes place at constant temperature. At one point in the dilution process the affinity of the surfactant becomes equal for both the phases that is it exhibits both hydrophilic and lipophilic properties. This point is known as the emulsion inversion point [4]. At this stage a layer of microemulsions is formed. A slight change in the proportion of oil and water causes instability of the microemulsion layer which disintegrates to form nanoemulsion that are kinetically stable. It was found out that with further

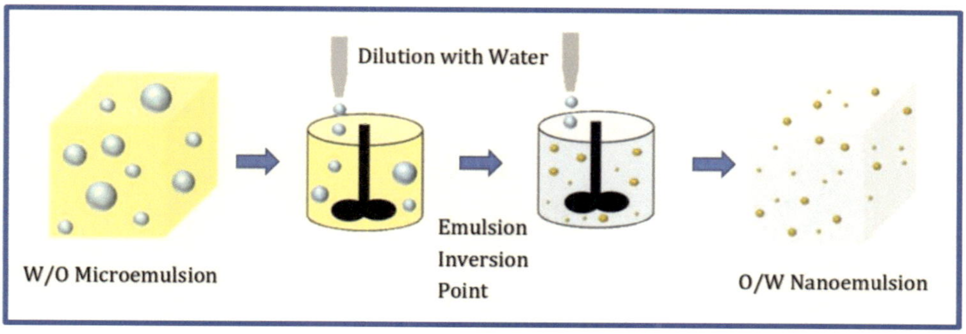

Figure 5.
Shows the process of phase inversion by dilution with aqueous phase [4].

dilution the droplet size does not change [28]. The properties of the nanoemulsion obtained depends on conditions such as shear rate and the addition rate [30]. Heat sensitive compounds can be encapsulated by using this fabrication method for nanoemulsions.

To obtain an O/W nanoemulsion, initially a W/O microemulsion (consisting of surfactant) is required to which aqueous phase is added in a controlled manner. The resulting system is stirred for breakdown of the residues and for homogeneity [24]. This process is depicted in **Figure 5**.

3. Applications of nanoemulsions

3.1 Drug delivery

A great attention is given towards the use of nanoemulsions in research, dosage form design and pharmacotherapy owing to their optical clarity, ease of preparation, thermodynamic stability and increased surface area. Some of the problems associated with conventional drug delivery systems such as low bioavailability and noncompliance which can be overcome by nanoemulsions are discussed here (**Figure 6**).

3.1.1 Parenteral delivery

It is the most effective form of drug delivery system usually adopted for active ingredients with low bioavailability and narrow therapeutic index. Here the therapeutic peptides or drugs prepared in the form of solutions or suspensions, are given as injections. Intravenous, intramuscular and subcutaneous drug delivery systems are the most commonly used parenteral routes.

Dissolution of enormous amounts of hydrophobic compounds coupled with mutual compatibility and ability to safeguard drugs from hydrolysis and enzymatic degradation make nanoemulsions ideal vehicles for parenteral transport [32]. Nanoemulsions help in sustained and controlled drug delivery through parenteral routes. Since nano emulsions are cleared more slowly (more residence time) than the coarse particles, they are advantageous over macroemulsion systems when delivered parenterally [33].

Nanoemulsions loaded with thalidomide have been synthesized. A dose as low as 25 mg leads to plasma concentrations which can be therapeutic when delivered through this system [34].

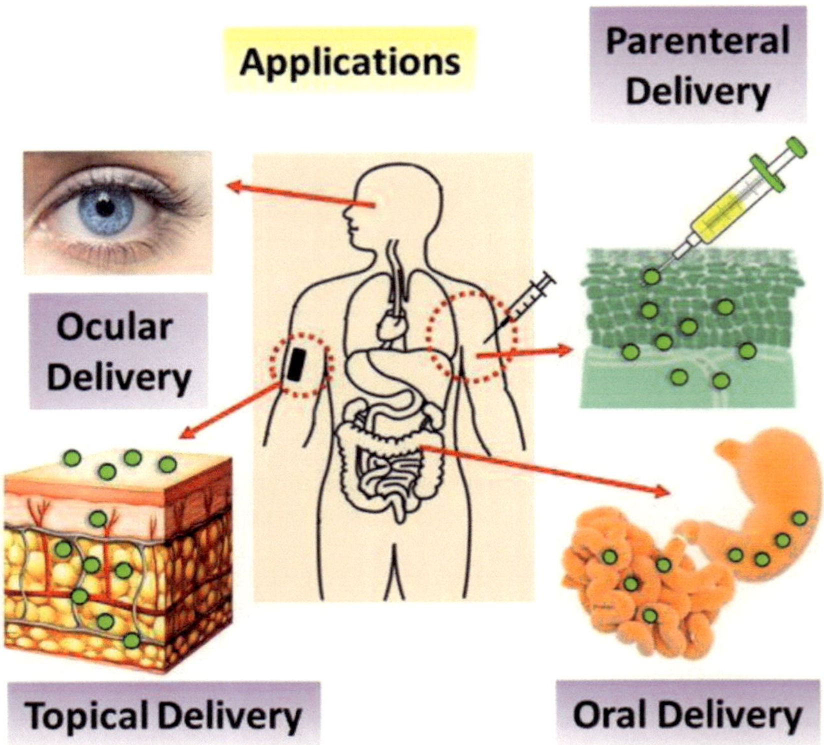

Figure 6.
Applications of nanoemulsions in drug delivery [31].

3.1.2 Oral and topical drug delivery systems

Owing to patient compliance, convenience, ease of formulation and higher absorption in the intestine, oral drug delivery is the most widely distributed and preferred form of drug administration.

When compared to conventional oral formulations, nano emulsion formulations provide several benefits in oral drug administration. Some of these benefits include increased absorption, improved clinical potency and decreased drug toxicity [33]. Hence, drugs such as steroids, hormones, diuretics and antibiotics can be ideally delivered using nano emulsions.

Topical drug delivery also has several advantages over other modes of drug administration such as the avoidance of hepatic first pass metabolism, reduction of toxicity and targeted drug delivery to the affected portion of the skin. Here self-administration is also possible. The transparent nature and fluidity of nanoemulsions not only gives a pleasant skin feel but also helps in eliminating the drug input by just removing the transdermal patch without any irritation [32].

Owing to the large surface area of the droplets, nanoemulsions enable rapid penetration of active ingredients through the skin. This is one of the most valuable properties of nanoemulsions due to which the use of special penetration enhancers which cause incompatibility of the formulation can be minimized [33].

3.1.3 Ocular and pulmonary drug delivery

For the treatment of eye diseases, drugs are delivered topically in ocular dosage forms such as solutions, suspensions and ointments. Due to physiologically protective mechanisms such as tear dilution, lacrimal drainage, protein binding and enzymatic degradation which are activated as soon as the ophthalmic solutions of the drug are applied, typically less than 3% of these topically applied drugs permeate the corneal epithelium, reach the aqueous humor and finally enter the systemic circulation. O/W Nanoemulsions (Oil in Water nano emulsions) have been researched for ocular administration to dissolve poorly soluble drugs, to increase absorption and also to attain prolong release profile [33].

Cationic submicron emulsions are promising carriers for DNA vaccines to the lung since they are able to transfect pulmonary epithelial cells thereby inducing cross priming of antigen-presenting cells and directly activate dendritic cells, resulting in stimulation of antigen-specific T-cells [35].

There might be adverse side effects of oils and surfactants on the alveoli of lungs. Hence extensive studies are required for the development of successful inhalable submicron emulsions for pulmonary delivery [32].

3.2 Nanoemulsions in biotechnology

Nanoemulsions serve as a waterproof medium for bio-catalytic or enzymatic reactions to occur. The enzymes in low water content exhibit greater solubility in non-polar reactants and higher thermal stability. As a result, the thermodynamic equilibrium also shifts towards condensation [36].

Reactions such as synthesis of esters, peptides and sugar acetals transesterification, hydrolysis and steroid transformation are catalyzed by enzymes in nanoemulsions. Lipases are the most widely used class of enzymes in microemulsion-based reactions [33].

3.3 Nanoemulsions as non-toxic disinfectant cleaners

The non-toxic disinfectant cleaner developed by Enviro Systems Inc. has wide applications in commercial markets such as healthcare, hospitality, travel, food processing, and military. This product which kills a wide spectrum of bacteria, virus and fungi in 5–10 minutes without any hazards needs no warning labels. This can be absorbed through the skin, inhaled or swallowed without causing any irritation to the eyes and does not cause any harmful effects [33]. One such NE is Parachlorometaxylenol (PCMX) marketed as EcoTru [36].

3.4 Nanoemulsions in the food sector

Lipophilic compounds such as flavors, omega3fatty acids, vitamins, nutraceuticals and preservatives can be encapsulated, stabilized and delivered using nanoemulsions. This is one of the emerging fields in the food industry [37]. Research is mainly focused on nanoemulsion technology that is suited for functional foods.

3.4.1 Encapsulation of lipophilic components using nano-emulsions

Encapsulation is a useful technique to deliver bioactive molecules within living cells. Here, the bioactive ingredient is entrapped in a core or filled within a carrier (coating, matrix, membrane, capsule, or shell) [38].

This technology is used in food industry to:

- Mask the unpleasant taste or odor of some bioactive materials.

- Increase the bioavailability of some components.

- Improve stability of food ingredients.
- Decrease air-induced food degradation.
- Reduce Evaporation of food aroma.

Another important application of this technique that is worth mentioning is in probiotics [38]. (Probiotics are the microorganisms that provide health benefits when consumed in adequate amounts) (**Figure 7**).

Biologically active lipids such as omega-3 fatty acids can be encapsulated by nano-emulsions from food grade ingredients. Omega-3 fatty acid supplementation has a protective effect against cancer, cardiac death, sudden death, cognitive aging, asthma, inflammation and myocardial infarction [38]. α-Linolenic acid (ALA), an Omega-3 fatty acid, is one of two essential fatty acids together with linoleic acid. ALA is necessary for health and cannot be synthesized within the human body [38].

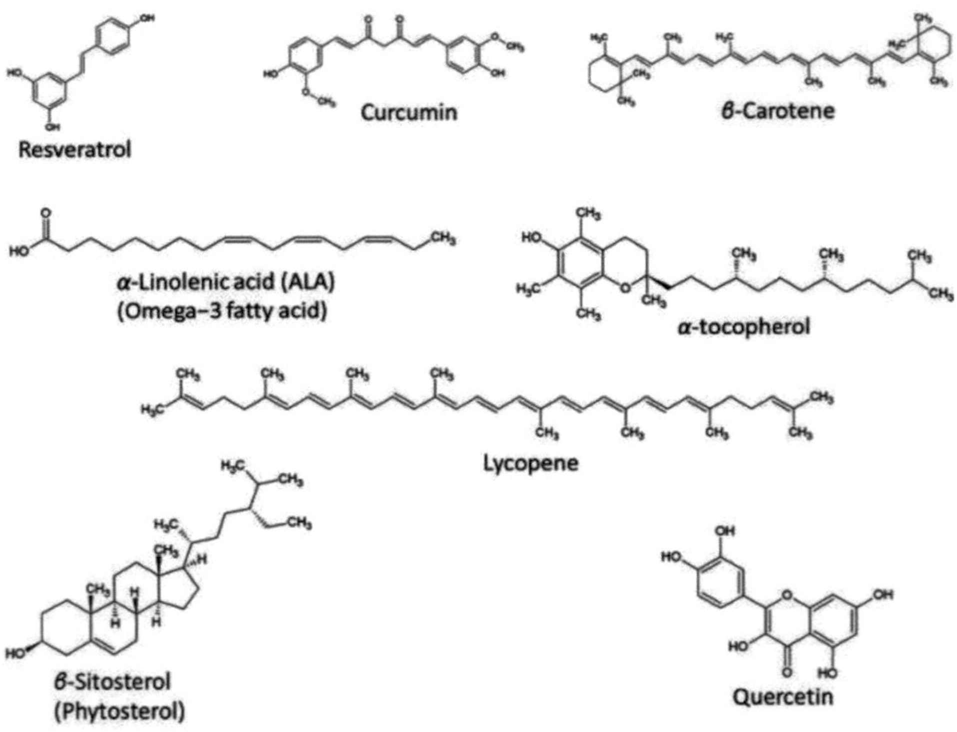

Figure 7.
Applications of nano-emulsion based delivery systems in food industry [38].

3.5 Nanoemulsions in cosmetics

What are cosmetics? *'Personal Care Products' or 'Beauty Products' such as skin creams, lotions, perfumes, lipsticks, fingernail polishes, eye and face make-up products, hair dyes and deodorants which are used by people to cleanse or change the look of the face or body are cosmetics. They do not alter the body's structure and function. Cosmetics are superficial and can also be therapeutic.*

Nanoemulsions are potential vehicles for optimum dispersion of active ingredients and their controlled delivery in particular skin layers. Their lipophilic interior makes them more suitable for the transport of lipophilic substances when compared to liposomes. Small sized droplets and high surface area of nano-emulsions contribute greatly in inhibiting inherent creaming, flocculation, sedimentation and coalescence which are frequently observed in macroemulsions [33]. They have gained popularity in cosmetics due to ease of permeability and penetration into the skin owing to their small size and high surface area, solubility, transparency and color.

Advantages of using Nanoemulsions in cosmetics

- As mentioned earlier, nano-emulsions help to overcome problems such as inherent creaming, flocculation, coalescence, and sedimentation.

- Relatively less surfactants (5–10%) [39] are used for nano-emulsion formulation (approved for human use) making them non-toxic and non-irritant. Hence, they can be easily applied onto the skin and mucous membranes [40].

- Healthy human and animal cells are not damaged by nano-emulsions. As a result, they are best suited for human and veterinary therapeutic purposes [40].

- Nano-emulsions can also be easily formulated in the form of foams, cream, liquids, sprays.

- Significant improvement in dry hair aspect (after several shampoos) is obtained with a prolonged effect after a cationic nanoemulsion use [39].

Oil in water nanoemulsions, due to their lipophilic interior have numerous applications in cosmetics such as formulation of lotions, sunscreens and skin creams, hair care products and make up removal substances. Other extremely important and fast growing applications include Emulsion based wet wipes and PEG (Polyethylene glycol) free nanoemulsions [40].

3.5.1 PEG free nanoemulsions and emulsion based wet wipes

The recent trends are shifting towards highly effective safer and natural cosmetic products. Cosmetic manufacturers are developing new methods to prepare nano versions of formulations for better permeability, effectiveness and increased customer satisfaction. This new technology is based on manufacturing low viscosity oil in water nanoemulsions which are free of synthetic chemicals like PEG. The phase inversion method of fabricating nanoemulsions leads to an important application in the cosmetic industry -formulating lotions to be impregnated in wet wipes. These are primarily used for make-up removal and wet wipes [11].

Industries are working on developing low energy methods coupled with low input homogenizers and absence of PEG, energy input for heating/cooling steps to formulate natural products [11]. The main constituents of this formulation are PEG free emulsifiers, cosmetic oils in high amounts and co-surfactants. When water is added to such a liquid and clear oil phase, a temporary micro-emulsion is formed which is then converted to stable, low-viscosity nanoemulsion. This low viscous phase is beneficial as it helps in converting the emulsion from water in oil to oil in water nanoemulsion. This step being dependent on water concentration is termed as phase inversion concentration which occurs by elimination of the co-surfactant into water. Some of these products developed are TEGO Wipe DE and TEGO Wipe DE PF which are based on nanoemulsions impregnated in wet-wipes and also used for lotions and sprays [41, 42].

3.5.1.1 Skin as a barrier and penetration pathways through the skin

The human body is protected from external harm such as chemical and micro-organism intrusion, UV exposure, dryness and mechanical damage by a natural barrier, the skin.

Hence the multi-layered skin serves as the first line of body defense. The external and internal layers of the skin mainly include the stratum corneum (composed of dead keratinized cells) and below the stratum corneum are the epidermis, dermis and the subcutaneous tissue. The excellent barrier properties of the skin are due to the presence of lipid matrix (containing ceramides, fatty acids, cholesterol and cholesteryl esters) among the keratinized cells which has cement like property [28].

When a drug or an active ingredient is topically applied on the skin surface, there theoretically three different ways through which the drug/active ingredient can penetrate through the skin. These are known as 'The Penetration Pathways'. The penetration pathways are:

- The intercellular pathway
- The hair follicle pathway
- The transcellular pathway

The intercellular pathway: This is the most widely known pathway. In this pathway, diffusion of the active ingredient occurs through the stratum corneum via the lipid layers surrounding the corneocytes.

The hair follicle pathway: The hair follicles are surrounded by a dense network of blood capillaries which support efficient penetration. These serve as reservoir for the topically applied active compound.

The transcellular pathway: This is the less understood pathway. Here the drugs are directly transported through the lipid layers and corneocytes to the living cells (**Figure 8**).

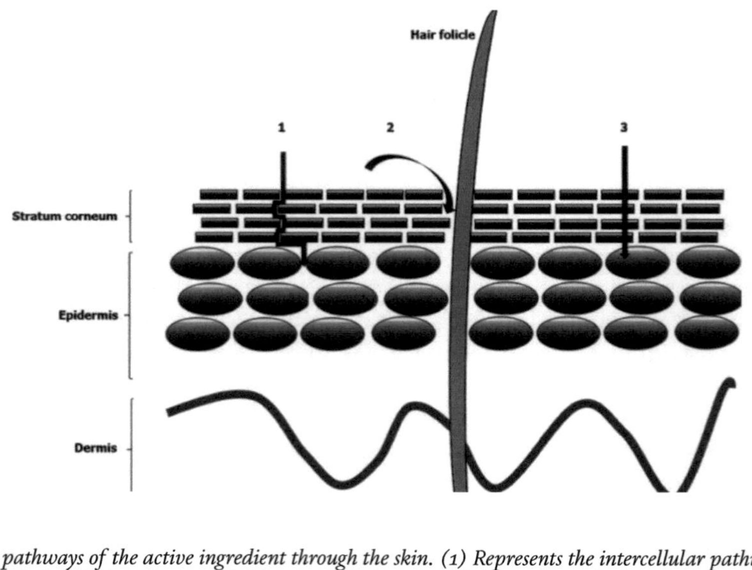

Figure 8.
Penetration pathways of the active ingredient through the skin. (1) Represents the intercellular pathway, (2) represents the hair follicle pathway and (3) represents the transcellular pathway [28].

3.5.2 Nanogels

While nanoemulsions are an efficient vehicle for transferring and administering topical drugs into the body, they have some limitations. These are low viscosity, spreadability constraints due to skin as a barrier and rheological properties.

To overcome these problems associated with nanoemulsions, a hydrophilic gelling system was integrated with nanoemulsions to increase the efficiency of transdermal drug delivery. These integrated systems are termed as nano emulgels [43].

Nanoemulgel are usually three-dimensional, spherical gels composed of a cross-linked network of polymeric (natural or synthetic substances) [44]. They are highly preferred over other nanomaterials for drug delivery due to their unique and advantageous features. The most important ones being biocompatibility, stimuli–response behavior softness, their ability to swell up to achieve a controlled, triggered response at the target site. They also protect the guest molecules i.e., the molecules they are carrying from degradation and elimination.

The versatility of their architecture allows for incorporation of a plethora of guest molecules ranging from inorganic nanoparticles to biomacromolecules like proteins and DNA with suitable modifications of the materials used for their construction without compromising their gel behavior. This multi-functionality and stability is hard to find in other types of nanoparticulate systems [43, 45, 46].

Nanogel properties can be used in various fields to achieve biomedical applications.

Stimuli response behavior: This involves the response of the nanogels to the external environment in the body such as pH, temperature, redox reactions, enzyme concentration etc. It employs the unique ability of the gel network to swell and unwell for this purpose. The nanogels can be composed of different materials depending on the type of response to be initiated. The deswelling and swelling occurs in the presence of changes in pH and concentration of the surrounding environment. For example, a gel network made of polysaccharide functionalized with PBA (aminophenyl boronic acid) is used to detect the fluctuations in glucose concentration, pH, concentration of the cationic and anionic groups bound to the gel and the PBA grafted to the gel which induces the release insulin by deswelling [44].

Another very important feature is protecting the cargo molecules from degradation and elimination and early clearance carrying small molecules for drug delivery by retaining them within the gel via hydrogen-bonding and hydrophobic interactions. Cationic gel polymers are also useful in carrying molecules of opposite charges such as oligonucleotides, proteins, RNA molecules or a combination of them to achieve multi-target drug delivery. This has been proven to be useful for cancer treatment in animals.

These hydrogels show immense amount of versatility, biocompatibility and fluid like transport properties which make them ideal carriers for imaging probes and scanning techniques such as optical imaging and multi-modal scanning [46]. It is also useful for anti-aging, skin care and moisturizing creams [31].

4. Conclusions

Nanoemulsions are a relatively new class of dispersions which have gained popularity due to their high efficiency in delivery. There have been a lot of efforts and research to develop the preparation methods of nanoemulsions. This emerging component of nanotechnology has become an irreplaceable part and parcel of the cosmetic and pharmaceutical industries. Further research and development in this field can prove to be very crucial for these industries. Nanoemulsions have a huge potential to change the approaches to many fields as discussed in this chapter and it can also play much more important roles in the future due to its uniqueness and increasing research in its field.

5. Future scope

The market for nanoemulsions is expected to grow at the rate of 8.8% between the years 2018 and 2023 [47].The growth is expected to be driven by the increasing demand for the treatment of chronic diseases and vaccine development. Nanoemulsions have huge potential to improve the efficacy of cancer immunotherapy by multiple folds. The major hindrance to the growth is the expensive manufacturing methods and scaling up with cost effectiveness. The technological innovation and scaling opportunities is expected to decrease the cost of production. Due to its unique properties, nanoemulsions can also become an active component in the chemical, agricultural and engineering fields. Nanoemulsions can also find an application as drug delivery platforms to noval phytopharmaceuticals given the new interest in herbal drug formulations in the world [32].

Acknowledgements

The authors listed in this paper wish to express their appreciation to the RSST trust Bangalore for their continuation support and encouragement. As a corresponding author, I also express my sincere thanks to all other authors whose valuable contribution and important comments make this manuscript in this form.

Conflict of interest

The authors listed in this paper have no conflict of interest known best from our side. There was also no problem related to funding. All authors have contributed equally with their valuable comments which made the manuscript to this form.

Funding information

There was no funding provided for the above research and preparation of the manuscript.

Author details

Praveen Kumar Gupta*, Nividha Bhandari, Hardik N. Shah, Vartika Khanchandani, R. Keerthana, Vidhyavathy Nagarajan and Lingayya Hiremath
Department of Biotechnology, RV College of Engineering, Bangalore, India

*Address all correspondence to: praveenkgupta@rvce.edu.in

References

[1] Tadros T. Emulsion systems. In: Tadros T, editor. Encyclopedia of Colloida and Interface Science. Berlin, Heidelberg: Springer; 2013. Available from: https://www.springer.com/in/book/9783642206641

[2] IUPAC. Compendium of polymer terminology and nomenclature. In: Jones RG, Kahovec J, Stepto R, Wilks ES, Hess M, Kitayama T, Metanomski WV, editors. IUPAC Recommendations 2008 (the "Purple Book"). Cambridge, UK: RSC Publishing; 2008

[3] Helmenstine AM. Surface Tension Definition and Causes. New York: ThoughtCo; 28 Dec 2018. Available from: https://www.thoughtco.com/definition-of-surface-tension-in-chemistry-605713

[4] Ankur G, Burak E, Hatton T, Patrick SD. Nanoemulsions: Formulation, properties and applications. Soft Matter. 2016;**12**(11):2826-2841. DOI : 10.1039/C5SM02958A

[5] Jasima H, Džana O, Alisa E, Edina V, Ognjenka R. Preperation of nanoemulsions by high-energy and lowenergy emulsification methods. In: Badnjevic A, editor. CMBEBIH 2017. IFMBE Proceeding, Vol. 62. Singapore: Springer; 2017

[6] Nguyen TTL, Anton N, Vandamme TF. Oral pellets loaded with nanoemulsions. In: Andronescu E, Grumezescu AM, editors. Micro and Nano Technologies, Nanostructures for Oral Medicine. Amsterdam: Elsevier; 2017. pp. 203-230. ISBN: 9780323477208

[7] Verma G, Rajagopalan MD, Valluru R, Sridhar KA. Nanoparticles: A novel approach to target tumors. In: Grumezescu AM, editor. Nano- and Microscale Drug Delivery Systems. Amsterdam, Netherlands: Elsevier; 2017. pp. 113-129. ISBN: 9780323527279

[8] Bharathi Priya L, Baskaran R, Vijaya Padma V. Phytonanoconjugates in oral medicine. In: Andronescu E, Grumezescu AM, editors. Micro and Nano Technologies, Nanostructures for Oral Medicine. Amsterdam, Netherlands: Elsevier; 2017. pp. 639-668. ISBN: 9780323477208

[9] Setya S, Talegaonkar S, Razdan BK. Nanoemulsions: Formulation methods and stability aspects. World Journal of Pharmacy and Pharmaceutical Sciences. 2014;**3**(2):2214-2228. ISSN: 2278-4357

[10] Guo H, Wilking JN, Liang D, Mason TG, Harden JL. Leheny RL. Slow nondiffusive dynamics in concentrated nanoemulsions Physical Review. E, Statistical, Nonlinear, and Soft Matter Physics. American Physical Society. 2007;**75**:041407. DOI: 10.1103/PhysRevE.75.041401

[11] Mishra RK, Soni GC, Mishra RP. A review article: On nanoemulsion. World Journal of Pharmacy and Pharmaceutical Sciences. 2014;**3**(9):258-274. ISSN: 2278-4357

[12] Pengon S, Chinatangkul N, Limmatvapirat C, Limmatvapirat S. The effect of surfactant on the physical properties of coconut oil nanoemulsions. Asian Journal of Pharmaceutical Sciences. 2018;**13**(5):409-414. ISSN: 1818-0876

[13] Badruddoza AZM, Gupta A, Myerson AS, Trout BL, Doyle PS. Low energy nanoemulsions as templates for the formulation of hydrophobic drugs. Advanced Theraputics. 2018;**1**(1):1700020-28

[14] Rodrigues IA, Ramos A d S, Falcão DQ, et al. Development of nanoemulsions to enhance the antileishmanial activity of copaifera pauper oeoresins. BioMed Research

International. 2018;**2018**: Article ID 9781724, 9 p

[15] Jin KD, Hassan B, Sushant A. Synthesizing pickering nanoemulsions by vapor condensation. ACS Applied Materials and Interfaces. 2018;**10**:21746-21754. DOI: 10.1021/acsami.8b06467

[16] Jintapattanakit A. Preparation of nanoemulsions by phase inversion temperature (PIT) method. Pharmaceutical Sciences Asia. 2018;**45**(1):1-12. DOI: 10.29090/psa.2018.01.001

[17] Jennifer K. Optimization of the fabrication, stability, and performance of food grade nanoemulsions with low and high energy methods [doctoral dissertations]. 582. 2016. Available from: https://scholarworks.umass.edu/dissertations_2/582

[18] Tadros T et al. Formation and stability of nano-emulsions. Advances in Colloid and Interface Science. 2004;**108**:303-318. DOI: 10.1021/la010808c

[19] Maa YF, Hsu C. Performance of sonication and microfluidization for liquid–liquid emulsification. Pharmaceutical Development and Technology. 1999;**4**(2):233. DOI: 10.1081/PDT-100101357

[20] Gupta PK, Pandit JK, Kumar A, Swaroop P, Gupta S. Pharmaceutical nanotechnology novel nanoemulsion–high energy emulsification preparation, evaluation and application. The Pharma Research. 2010;**3**:117-138. ISSN: 0975-8216

[21] Benichou A, Aserin A, Garti N. Polyols, high pressure, and refractive indices equalization for improved stability of W/O emulsions for food applications. Journal of Dispersion Science and Technology. 2001;**22**(2/3):269-280. DOI: 10.1081/DIS-100105214

[22] McClements DJ. Food Emulsions: Principles, Practices, and Techniques. Washington, DC: CRC Press INC.; 2005. ISBN: 0-8493-2023-2

[23] McClements DJ, Rao J. Food-grade nanoemulsions: Formulation, fabrication, properties, performance, biological fate, and potential toxicity. Critical Reviews in Food Science and Nutrition. 2011;**51**:285-330. DOI: 10.1080/10408398.2011.559558

[24] Komaiko JS, McClements DJ. Formation of food-grade Nanoemulsions using low-energy preparation methods: A review of available methods. Comprehensive Reviews in Food Science and Food Safety. 2016;**15**(2):331-352. DOI: 10.1111/1541-4337.12189

[25] Abhijit A Date, Neha Desai, Rahul Dixit, Mangal Nagarsenker. Self-nanoemulsifying drug delivery systems: Formulation insights, applications and advances. Nanomedicine. 2010;**5**(10):1595-1616. DOI: 10.2217/nnm.10.126

[26] Kotta S, Khan AW, Ansari SH, Sharma RK, Ali J. Formulation of nanoemulsion: A comparison between phase inversion composition method and high-pressure homogenization method. Drug Delivery. 2015;**22**(4):455-466. DOI: 10.3109/10717544.2013.866992

[27] Garcia-Celma MJ, Homs M, Morales D, Solans C. Nano-emulsions for pharmaceutical applications. In: Sanchez-Dominguez M, Rodriguez-Abreu C, editors. Nanocolloids: A Meeting Point for Scientists and Technologists, Chapter 11. Rockville: Elsevier; 2016. pp. 365-388. DOI: 10.01016/B978-0-12-801578-0.00011-4

[28] Yukuyama MN, Ghisleni DDM, Pinto TJA, Bou-Chacra NA. Nanoemulsion: Process selection and application in cosmetics—A review.

International Journal of Cosmetic Science. 2016;**38**:13-24. DOI: 10.1111/ics.12260

[29] Jaiswal M, Dudhe R, Sharma PK. Nanoemulsion: An advanced mode of drug delivery system. 3 Biotech. 2015;**5**(2):123-127. DOI: 10.1007/s13205-014-0214-0

[30] Solè I, Pey CM, Maestro A, González C, Porras M, Solans C, et al. Nano-emulsions prepared by the phase inversion composition method: Preparation variables and scale up. Journal of Colloid and Interface Science. 2010;**344**:417-423. DOI: 10.1016/j.jcis.2009.11.046

[31] Singh Y, Meher JG, et al. Nanoemulsion: Concepts, development and applications in drug delivery. Journal of Controlled Release. 2017;**252**:28-49

[32] Lovelyn C, Attama AA. Current state of nanoemulsions in drug delivery. Journal of Biomaterials and Nanobiotechnology. 2011;**2**:626-639. Available from: http://www.scirp.org/journal/jbnb

[33] Ozgun S. Nanoemulsions in Cosmetics. Project Report. Material Science and Engineering Department, Anadolu Universitesi. Turkey: Eskişehir; 2013. pp. 8-9

[34] Patel R, Patel KP. Advances in novel parenteral drug delivery systems. Asian Journal of Pharmaceutics. 2010;**4**(3):193-199. DOI: 10.4103/0973-8398.72117

[35] Bivas-Benita M, Oudshoorn M, Romeijn S, et al. Cationic submicron emulsions for pulmonary DNA immunization. Journal of Controlled Release. 2004;**100**(1):145-155. DOI: 10.1016/j.jconrel.2004.08.008

[36] Chavda VP, Shah D. A review on novel emulsification technique: A nanoemulsion. Journal of Pharmacology and Toxicological Studies. 2017;**5**(1):32-33

[37] Nethaji DK, Parambil KA. Development and applications of nano emulsion in food technology. International Journal of Science, Engineering and Management. 2017;**2**(12):60-61

[38] Salem M, Ezzat S. Nanoemulsions in Food Industry. 2018. Available from: https://www.intechopen.com/online-first/nanoemulsions-in-food-industry/. DOI: 10.5772/intechopen.79447

[39] Chellapa P, Ariffin FD, Eid AM, Almahgoubi1 AA, Mohamed1 AT, Issa YS, et al. Nanoemulsion for cosmetic application. European Journal of Biomedical and Pharmaceutical Sciences. 2016;**3**(7):8-11

[40] Sharma S, Sarangdevot K. Nanoemulsions for cosmetics. International Journal of Advanced Research in Pharmaceutical and Bio Sciences. 2012;**2**(3):408-415

[41] Friedrich A, Biehl P, Zimmermann S, Trambitas A, Meyer J. Novel multipurpose PEG-free W/O silicone emulsifier with performance benefits in color cosmetics. SOFW Journals. 2013

[42] Meyer J. Nanoemulsions for PEG free cosmetics by simple dilution

[43] Arora R, Aggarwal G, Harikumar SL, Kaur K. Nanoemulsions based hydrogel for enhanced transdermal delivery of Ketoprofen. Advances in Pharmaceutics. 2014;**2014**

[44] Soni KS, Desale SS, Bronich TK. Nanogels: An overview of properties, biomedical applications and obstacles for clinical translation. 2015. DOI: 10.1016/j.jconrel.2015.11.009

[45] Chellapa P, Mohamed AT, Keleb EI, Elmahgoubi A, Eid AM, Issa YS, et al.

Nanoemulsions and nanoemulgels as a topical formulation. IOSR Journal of Pharmacy. 2015;5

[46] More A, Ambekar AW. Development and characterization of nanoemulsion gel for topical drug delivery of nabumetone. Human Journals. 2016;7

[47] Nanoemulsion Market Share—Segmented by application in Drug Class (Anesthetics, Antibiotics, NSAIDs, immunosuppressants, Steroids), Route of Administration, Distribution Channel, and Geography—Growth, Trends, and Forecast (2018-2023). Available from: https://www.mordorintelligence.com/industry-reports/nanoemulsion-market

Section 4

Applications

Chapter 6

Importance of Surface Energy in Nanoemulsion

Kaustav Bhattacharjee

Abstract

The emerging prospects of nanoscience and nanotechnology have an enormous promise to revolutionize various aspects of human life. In this context, the application of nanoemulsion stands at the vanguard of introducing newer dimensions to the way we see the everyday world. Naturally, the preparation and stability of nanoemulsion demand a precise understanding of the underlying forces of interaction toward achieving a greater control over their functionality and regulating them. The stability of nanoemulsion is primarily governed by the conjugate and complex interplay of van der Waals forces and steric interactions. The present chapter will be dedicated to the discussion of the regulatory roles of these forces in dictating the stability of nanoemulsion with particular emphasis on the origin of these fundamental forces from a molecular-level viewpoint.

Keywords: surface tension, curvature, Ostwald ripening, steric stabilization

1. Introduction

An emulsion is a special form of colloid where one liquid (dispersed phase) is uniformly distributed in another liquid (dispersion matrix) [1, 2]. Minute droplets of dispersed phase which is otherwise immiscible in the dispersion medium form a statistically distributed system encapsulated within the matrix having a boundary between the phases, called the "interface." Oil (O), water (W) and surfactants (S) are the three key components of the emulsion. Depending on the order of preparation, it can be water in oil (W/O), oil in water (O/W), water in oil in water (W/O/W), or oil in water in oil (O/W/O) [1]. In view of the interest of the present discussion, we will focus only on O/W emulsion throughout the chapter. O/W emulsion consists of small globular compartments composed of oil (lipophilic) and surfactant which can be conveniently dispersed in water.

Such system has a great deal of research interest in the field of food, pharmaceutical, agrochemical and other industries with certain benefits ([3–5] and references therein). *First*, they usually have improved stability to particle aggregation and eventual sedimentation. *Second*, by virtue of the appropriate small size of the droplets, they only weakly scatter light wave and therefore are advantageous for adding values into the products that need to be optically transparent (or only slightly turbid). *Third*, they can be tailor made to have novel rheological properties. *Fourth*, they have a much higher bioavailability of specific types of bioactive molecules contained within the dispersed phase.

Due to the fact that the environment of a molecule at an interface is different than those of the bulk phase, an interface is always associated with surface free energy. The free energy per unit area, measured in terms of the surface tension (γ), is the minimum amount of work required to create a new area of that interface [6]. Minimization of the interfacial contact area is therefore a spontaneous process. A surface-active agent (or surfactant) is a substance which at low concentration can adsorb at the interface, thereby reducing the amount of work required to expand it [7, 8]. In general, surfactants are amphiphilic molecules which reduce the interfacial surface tension due to their dual chemical nature and strong tendency to self-assemble above a certain concentration (precisely a narrow concentration range) at a given temperature, known as the critical micellar concentration (CMC). There are two most common types of emulsion classified in the literature based on their stability and structural components, namely, *microemulsion* and *nanoemulsion* [3]. Both types have droplet radius in the range of sub-100 nanometers. Despite several dissimilarities between these two kinds, it has been unfortunate that there has been a great confusion and widespread errors in their usage in the scientific literature. The confusion comes from the prefixes used to denote them. As the terms "micro-" (10^{-6}) and "nano-" (10^{-9}) suggest, it is assumed that nanoemulsions contain droplets that are smaller than microemulsion. In practice, the opposite is often found; indeed, there has also been much disagreement about the critical droplet size to distinguish between the two. A clear distinction between these two types of liquid-in-liquid dispersion has been made in two recent studies [3, 4], and the interested readers are suggested to go through the references. However, a brief clarification is given in the present chapter.

A free energy diagram for the two systems is schematically shown in **Figure 1**. An O/W-type microemulsion is a thermodynamically stable isotropic dispersion of oil and surfactant in water. Nevertheless, it is strongly affected and even broken up by modulations in thermodynamic variables, such as temperature, composition, pH, etc. Microemulsions are formed spontaneously (without the need of an external agency) when surfactants are added to the oil-water mixture [3, 4]. The non-polar tails of surfactant molecules self-assemble to form a hydrophobic core where oil molecules can be stored and separated from the thermodynamically unfavorable aqueous phase of the surroundings. The final structure of such microenvironment may result in a spherelike (micelle or reverse micelle), cylinder-like (rod micelle), plane-like (lamellar micelle) or sponge-like (bicontinuous) shape.

On the other hand, an O/W-type nanoemulsion is a thermodynamically unstable isotropic dispersion of oil and surfactant in water [3, 4]. In principle, nanoemulsion

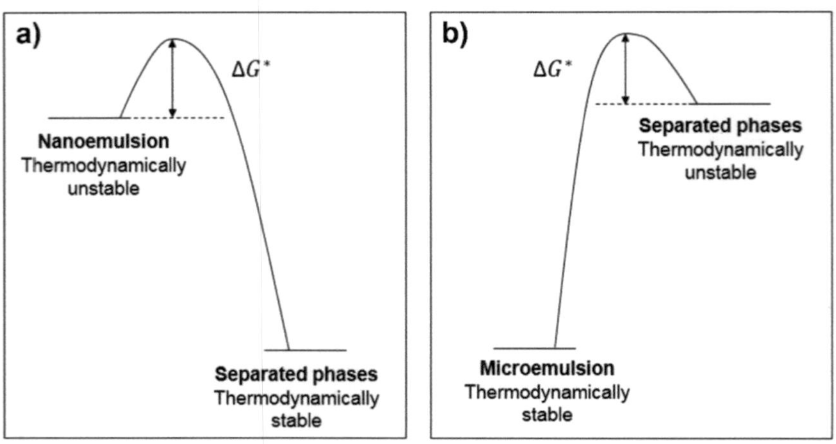

Figure 1.
Schematic diagram of the free energy of (a) nanoemulsion and (b) microemulsion system in comparison to their respective reference states. The two states are separated by an activation energy ΔG^*.

can be fabricated without addition of any surfactant molecule only by physical methods that involve energy input. However, such a system would be highly unstable with respect to droplet coalescence (merging of two droplets) and phase segregation. Therefore, surfactant is needed to ensure the kinetic stability of nanoemulsion during prolonged storage [8]. Nevertheless, under certain circumstances surfactant may impart negative effect on nanoemulsion stability, because of their ability to enhance the mass transfer processes which can cause significant change in droplet concentration, composition and size distribution [8]. The mass transport process is typically driven by differences in chemical potentials for the solutes in each microenvironment, and as a consequence, droplets tend to merge and transport the dissolved matter through the dispersion medium, by a process known as "Ostwald ripening." Another essential difference between micro- and nanoemulsion that is often neglected in the literature is the influence of the order in which different compounds are mixed together during preparation [3]. This point is particularly important for nanoemulsion. Nanoemulsions are only formed if the surfactants are first mixed with the oil phase and then the surfactant-oil mixture is added to the aqueous phase. If it is not followed, only a "macroscopic" emulsion will be generated. Microemulsion, on the contrary, will be strictly identical whatever the order in which the components are mixed (after equilibrium time).

The major advantages and disadvantages of nanoemulsion over microemulsion for the specific application purpose are summarized below [3, 4].

1.1 Advantages

(1) Due to their smaller droplet size, reduction under gravitational pull can be avoided in large extent, and, therefore, nanoemulsion never shows creaming and sedimentation problems, while these problems are quite common with conventional emulsion or even with microemulsion. With proper stabilization forces, nanoemulsions can be stored for a longer period than microemulsion.
(2) Nanoemulsions are very suitable for rapid penetration of active ingredients (pharmaceuticals and/or food) due to their smaller size and large surface area.
(3) Unlike microemulsion which requires high surfactant concentration (20% or higher), nanoemulsion can be formed using reasonably low surfactant concentration (5–10%).

1.2 Disadvantages

(1) Fabrication of nanoemulsion in many cases demands special and expensive instrumentation (high-pressure homogenizers or ultrasonics, microfluidizer, etc.), technique as well as higher concentration of surfactants. (2) The lacuna in the understanding of various fundamental issues associated with nanoemulsion strongly restricts its acceptability and applicability. Knowledge of proper interfacial chemistry, mechanism of Ostwald ripening and ingredients to overcome it are the key issues that need to be taken care of for the superior acceptability and applicability of nanoemulsion.

2. Thermodynamics and kinetics of nanoemulsion

2.1 Free energy diagram

The thermodynamic stability of a particular system is governed by the change in free energy between it and an appropriate reference state. Nanoemulsion is thermodynamically unstable, which means that the free energy of nanoemulsion is higher than the free energy of the separate phases (oil and water). For microemulsion, this condition is opposite. To understand the molecular basis of the free energy difference, let us consider that a system (nano-/microemulsion) exists in equilibrium between the initial and final states. The free energy change associated with the formation of the dispersion consists of an interfacial free energy term (ΔG_I) and a configuration entropy term ($-T\Delta S_{config}$) [9]:

$$\Delta G_{formation} = \Delta G_I - T\Delta S_{config}.$$

The free energy to increase the contact area (ΔA) at the interface is $\Delta G_I = \gamma\Delta A$, which is always positive; consequently, this term always opposes the formation of the dispersions. However, the interfacial tension (γ) depends on the curvature of the surfactant layer-decreasing as the curvature approaches to its optimum value. A phenomenological description of the dependence of interfacial tension on droplet curvature can be formulated as [10–12]:

$$\gamma = \gamma_0 + (\gamma_\infty + \gamma_0)\frac{(R_0 - R)^2}{R_0^2 + R^2},$$

where γ_∞ and γ_0 are, respectively, the interfacial tension values at the planar O/W interface and when the surfactant layer reaches its optimum curvature. R_0 is the droplet radius at the optimum curvature.

On the other hand, configuration entropy (ΔS_{config}) which depends on the number of arrangements accessible to the oil phase in an emulsified state is much greater than that in a non-emulsified state, and therefore it always favors the formation of the dispersion. An expression for the ΔS_{config} can be derived from the statistical analysis [9]:

$$\Delta S_{config} = -\frac{nk}{\phi}(\phi ln\phi + (1-\phi)\ln(1-\phi)),$$

where k is the Boltzmann constant, n is the number of droplets and ϕ is the disperse-phase volume fraction.

The plot for the different free energies with droplet radius assuming that the interfacial tension is the same as that for planer O/W interface is shown in **Figure 2(a)** [3]. It can be observed from the figure that the interfacial free energy contribution increases (hence unfavorable) with decrease in droplet size (increase in interfacial area), while the configuration of free energy becomes progressively negative (hence favorable) with decrease in droplet size (increase in number of different ways that droplet can be organized). The total free energy, however, becomes increasingly positive with decrease in droplet size, since the interfacial free energy term dominates the configuration entropy term.

This implies that the formation of nanoemulsion becomes increasingly thermodynamically unfavorable as the radius of the droplets fall and where the interfacial tension is similar to that at a planar surface. Interestingly, if the calculation were performed assuming the dependence of the interfacial tension on the curvature, the situation becomes more complex [3]. As shown in **Figure 2(b)**, with decrease in droplet radius from say, 1000 nm, there is an increase in interfacial free energy. But once the droplet becomes smaller below a certain radius, the interfacial free energy decreases and reaches a minimum value, before rising again with further decrease in radius. The interfacial free energy contribution still is the dominant factor, forcing the total free energy to attain a minimum value that is close to the optimum curvature of surfactant layer. Thus, under this consideration, one

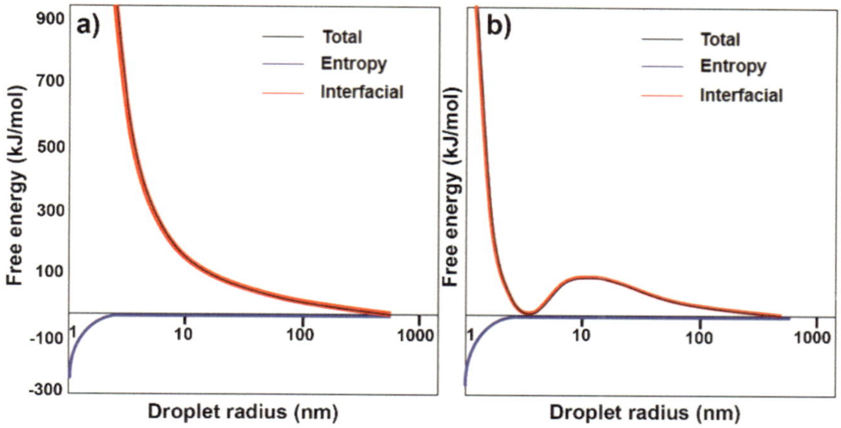

Figure 2.
Predicted variation of free energy change with droplet radius for the formation of emulsion state assuming (a) constant interfacial tension and (b) varying interfacial tension with curvature. Taken from Reference [3].

can assume that thermodynamically stable dispersion can be achieved when droplet radius is close to the optimum ($R = R_0$) and hence the interfacial free energy that otherwise opposes the emulsification can be appreciably reduced.

2.2 Kinetic picture

Kinetic stability, as to be contrasted with the thermodynamic stability, is determined by two important factors [3]: (1) Energy barriers: any energy barrier (or activation energy) that separates the two states (final and initial) will determine the rate of the conversion of one state to another. The height of this energy barrier depends on the forces operating in close proximity of two droplets, such as repulsive hydrodynamic and colloidal (steric and electrostatic) interactions [13]. Nanoemulsion can be made kinetically stable by introducing sufficiently large energy barrier (typically >20 kT) between the two states, while for microemulsion, there is still an activation energy required (in terms of mechanical agitation or heating the system) to reach the thermodynamically stable state after the components are brought into contact. (2) Mass transport phenomena: droplets in a nanoemulsion are particularly labile toward the growth over time by a process known as "Ostwald ripening" [8], in which solute molecules (or mass) are exchanged between the droplets via molecular diffusion through the solvent. Three alternative mechanisms have been proposed in the literature suggestive of that micelle which plays an important role in facilitating the mass exchange between the droplets by acting as carriers of oil molecules [8]. In mechanism *one*, oil molecules are transferred via direct micelle collisions, i.e., the rate is directly proportional to the volume fraction of micelle in solution. Numerous studies indicate a higher rate of mass transport above the CMC of the surfactant used. The lack of such CMC dependence for ionic surfactants however, may stem from the electrostatic repulsion between the droplet and micelle. In mechanism *two*, oil molecules exit the droplet, are exposed to the continuous phase and are soon captured by micelles in the immediate vicinity of the droplet. For non-ionic surfactant micelles, the higher rate of mass transfer is expected due to their higher solubilization capacity and the absence of electrostatic repulsion between the droplet and uncharged micelle. In mechanism *three*, a large number of oil molecules are released from the oil droplets collectively with excess of surfactant molecules to form a new micelle.

3. Formation of oil in water droplets from geometric and force balanced point of view

3.1 Role of surfactant in geometrical packing

We have already defined, in a limited sense, the word "surfactant" as an amphiphilic molecule which has the capacity to self-organize above a critical concentration. The process of self-assembly is dynamic in nature [13]. For those whose molecules at the air (or oil)-water interface are in exchange equilibrium with bulk solution are called soluble monolayer with a typical residence time on the order of 10^{-6} s. On the other hand, for those whose molecules are in less dynamic situation at the monolayer when the interface is expanded or compressed are called insoluble monolayer, and the time taken for such molecular exchange can vary typically from seconds to months (**Figure 3**).

The self-assembly of amphiphiles strongly depends on the two key surface forces that act at the interface: the "hydrophobic attraction" between the surfactant tails which induces the association of the molecules and the "hydrophilic repulsion" between the surfactant heads which helps them to remain in contact with water [13]. Though the selection of emulsifier in the preparation of either O/W or W/O nanoemulsions is still made on an empirical basis, however, a semiempirical scale is defined based on the relative percentage of hydrophilic to lipophilic groups in the surfactant molecules, known as hydrophilic-lipophilic balance (HLB number) [14], the HLB number is deduced from the preferential solubility of the surfactant in oil or water and HLB number can vary from 0 (very soluble in oil) to 20 (very soluble in water). Griffin postulated a simple equation to calculate the HLB number for non-ionic surfactant such as fatty acid ester [14], $HLB = 20 - \left(1 - \frac{S}{A}\right)$, where S is the saponification number and A is the acid number (**Figure 4**). Davies developed a method for calculating the HLB number for surfactants irrespective of their chemical nature, using empirically determined group numbers [15]: $HLB = 7 + \sum(hydrophilic\ group\ no) - \sum lipophilic\ group\ no$. Beerbower and Hills used the following expression for the HLB number [16]: $HLB = 20\left(\frac{M_H}{M_L + M_H}\right) = 20\left(\frac{V_H \rho_H}{V_L \rho_L + V_H \rho_H}\right)$, where M_H and M_L are the molecular weights of the hydrophilic and lipophilic portions of the surfactants. V_H and V_L are their corresponding molar volumes, whereas ρ_H and ρ_L are the respective densities. After adequately describing the interactions between the amphiphiles within an aggregate, we need to establish the

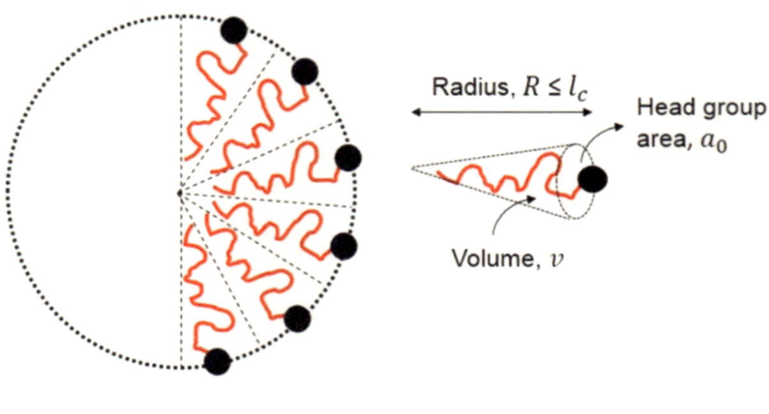

Figure 3.
Schematic representation of spherical micelle with illustration of different surfactant parameters. Packing factor = $\frac{V}{l_c a_0}$*. The chain volume, v, and chain length* l_c *set the aggregation limit.*

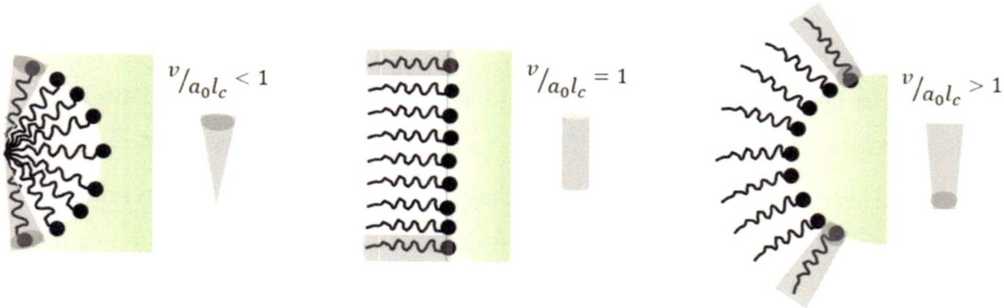

Figure 4.
View of the curvature of surfactant aggregates formed at various surfactant parameters, $\frac{v}{a_0 l_c}$*. For less than one condition (left), the interface curves toward chain region. For greater than one condition (right), the interface curves toward the polar region. When exactly equal to one (middle), the interface exhibits no preferential curvature.*

most favored structure of these aggregates. A convenient parameter to analyse such diverse structures is the dimensionless number, $v/a_0 l_c$, also known as "surfactant parameter" [17]. Close packing of amphiphiles leads to curved interfaces, and the direction of the curvature (either toward the polar or non-polar region) depends upon the value of this parameter. Here, l_c is a semiempirical parameter called the critical chain length of the same order (or somewhat less) as the fully extended molecular length of the chain, l_{max}; a_0 is the optimal head group area; and v is the volume of the hydrocarbon chain [17]. Once these parameters are specified for a given molecule, one may ascertain the most preferred geometrical packing. Gradation of the preferred structure with increasing surfactant parameter has been made as follows [13]: $v/a_0 l_c \le \frac{1}{3}$ for spherical to $\frac{1}{3} \le v/a_0 l_c \le \frac{1}{2}$ for ellipsoidal to $v/a_0 l_c \approx \frac{1}{2}$ for cylindrical or rodlike micelle to $\frac{1}{2} \le v/a_0 l_c \le 1$ for various interconnected structures or lameller phases to $v/a_0 l_c = 1$ for vesicles or extended bilayer and finally to a family of "inverted structure" for $v/a_0 l_c > 1$. Therefore, it can be concluded that in a non-condensed liquid phase the curvature is a function of the surfactant parameter at the liquid-liquid interface. *Here, we find surface tension at work*, and consequently it *results in a pressure imbalance* across a curved surface [6]. The origin of the tendency to minimize the surface energy of oil droplets in water is due to the imbalance in forces acting on a molecule at the interface compared to those acting in the bulk. From the basic fluid dynamics, we find that when a surface is curved there is a difference in pressure on the two sides of the surface which is described by the very important concept of "Laplace pressure" on the two sides of a curved (non-planer) surface, a spherical one being a special case of this in general [6].

3.2 Derivation of Laplace equation

Consider a spherical cap symmetric about z-axis. The pressure exerted on the curved interface by the two bulk phases will be different and will give rise to a force acting along the normal to the interface at each point on the curve. The cap will also feel a force arising from surface tension acting tangentially at all points on the curve. At mechanical equilibrium, these two forces cancel each other along the z-direction (**Figure 5**).

Force arising from the pressure difference: $F_z^{\Delta P} = (P^\alpha - P^\beta)\sum \delta A = (P^\alpha - P^\beta)\pi r_c^2$.

Force arising from surface tension: $F_z^\gamma = \frac{2\pi r_c^2 \gamma}{r}$.

At equilibrium the forces along z-direction: $F_z^{\Delta P} = F_z^{\gamma}$; or $(P^{\alpha} - P^{\beta})\pi r_c^2 = \frac{2\pi r_c^2 \gamma}{r}$.

So, $\Delta P = (P^{\alpha} - P^{\beta}) = \frac{2\gamma}{r}$, which is the *Laplace equation* for spherical surface.

For nonspherical interface, two orthogonal radii of curvature (r', r'') are needed, the *Laplace equation* then becomes $\Delta P = \gamma\left(\frac{1}{r'} + \frac{1}{r''}\right) = \frac{2\gamma}{r_m}$, where r_m is the mean curvature (inverse of radius) and is equal to $\frac{1}{r_m} = \frac{1}{2}\left(\frac{1}{r'} + \frac{1}{r''}\right)$. By our convention, if the interface encloses hydrophobic region, the mean curvature would be negative. Therefore, the Laplace pressure always drives the interface in the concave direction if the molecules are expanded or contracted. The high energy required for the formation of nanoemulsion droplets can be understood from the inverse relation of the pressure difference with the radius. The Laplace equation forms the theoretical basis of "Kelvin equation," which describes the effect of surface curvature of a liquid that changes with the equilibrium vapor pressure of the liquid [13]:

$\ln\left(\frac{P^c}{P^{\infty}}\right) = \left(\frac{2\gamma V_m}{r_m RT}\right)$, and we immediately obtain the *Kelvin radius*:

$r_k = \left(\frac{1}{r_1} + \frac{1}{r_2}\right)^{-1} = \frac{\gamma V_m}{RT\ln\left(\frac{P^c}{P^{\infty}}\right)}$, where P^c and P^{∞} are, respectively, the vapor pressures over the curved surface with mean curvature r_m and a flat surface $(r = \infty)$, V_m is the molar volume of the oil, γ is the interfacial surface tension, R is the gas constant and T is the absolute temperature.

3.3 The consequences of Kelvin equation are profound

One of them is to govern the process such as the growth of larger droplet in expanse of the smaller one in nanoemulsion. The vapor pressure of a spherical droplet will be greater than that of the same liquid with a flat surface. *The smaller the radius, the higher the vapor pressure* such that if there is a distribution of droplet size, the smaller one will tend to diminish, while the larger ones will tend to grow. The total energy of the two-phase system thus can be decreased via an increase in the

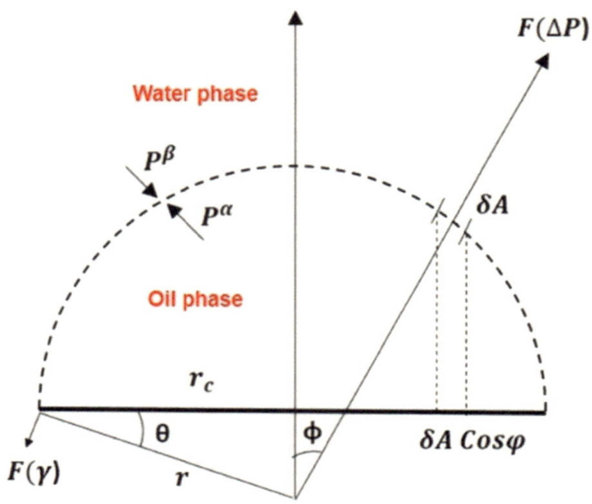

Figure 5.
Resolution of forces on spherical cap symmetrical about the z-axis and part of a spherical interface. The pressures exerted on the interface by the two bulk phases (α and β) will be different if the interface is curved, and this difference (ΔP) will give rise to a force acting along the normal to the interface at each point. The cap will also be subject to a force arising from surface tension acting tangentially at all points around the perimeter of the cap.

size scale of the second phase and accordingly a decrease in total interfacial area. Such a process is known as "Ostwald ripening." The driving force of the ripening process is the dependence of oil solubility on its size, as described by the Kelvin equation [18, 19]:

$$C(r) = C(\infty) \exp\left(\frac{2\gamma V_m}{RTr_m}\right) \approx C(\infty)\left(1 + \frac{2\gamma V_m}{RTr_m}\right),$$

where V_m is the molar volume of the oil and γ is the surface tension. This equation relates the solubility of droplet $C(r)$ with an arbitrary radius r to that of an infinite radius $C(\infty)$. The quantity $\left(\frac{2\pi V_m}{RTr_m}\right)$ has the dimension of length and is termed as the *capillary length* of the drop, typically on the order of ~1 nm. The Kelvin equation is derived from the dependence of chemical potential (μ) of the dispersed phase on its size by the relation [19]:

$$\Delta\mu(r) = \mu(r) - \mu(\infty) = \int_0^{2\gamma/r} V_m(p)dp \approx \frac{2\gamma V_m}{r}.$$

Another driving force for Ostwald ripening to occur in nanoemulsions is due to the polymorphic changes during redeposition of solute (such as drug molecules).

4. Theory of Ostwald ripening in nano-dispersion

So far, we came to know that the major instability of O/W nanoemulsions is caused due to the molecular exchange of oil molecules between droplets, a process known as *Ostwald ripening*. Although many specialists on emulsion stability are often unwilling to accept the concept of Ostwald ripening, instead, in most cases, they explain the destabilization under the framework of the more traditional mechanism—coalescence. Despite these disagreements in the literature, the theoretical development of the kinetic regime of the Ostwald ripening is a peculiar form of self-ordering and stimulates curiosity to the researchers till now. The detail mathematics of the theory of Ostwald ripening mechanism is not included in this chapter; rather, we will address a few selected topics of the theory. A more comprehensive discussion on this issue can be found in the reviews given in Refs. [19–21].

The major contribution in the theory of kinetics of Ostwald ripening in its contemporary form was initially formulated by Lifshitz and Slyozov [22] and then independently by Wagner [23] (known as LSW theory). Following publication of their findings, it became the seminal paper on which all subsequent theoretical works have been based. The theory is based on the following assumptions: (1) the particles (here droplets) of dispersed phase are spherical in shape, are fixed in space and are separated from each other by distance which are much larger than their sizes (true diluted system), (2) the mass transport is due only to the molecular diffusion through the dispersion medium and (3) the concentration of the molecularly dissolved species is constant except adjacent to the droplet boundaries.

4.1 Scaling the ripening problem

The ripening mechanism is characterized by two intervals, namely, the *transient* or short-time regime and the *asymptotic* limit or long-time regime. The transient limit is composed of a region of random variation of droplet size, whereas the asymptotic limit (t $\rightarrow \infty$) is ascribed to the region that shows a linear relationship between the cubes of the number average droplet radius ($r_N{}^3$) with time.

We follow the procedure used in reference [21] by Kabalnov et al. If $f(r,t)$ be the size (radius) distribution function of the polydisperse system, such that $f(r,t)dr$ is the number of particles per unit volume in the size range r to $r + dr$. The change in the distribution function with time can be expressed as

$$\frac{df(t,r)}{dt} = 0,$$

$$\text{or } \frac{\delta f}{\delta t} + \frac{\delta f}{\delta r}\frac{dr}{dt} = 0,$$

$$\text{or } \frac{\delta f}{\delta t} + \frac{\delta}{\delta r}(f\dot{r}) = 0,$$

where $\dot{r} = \frac{dr}{dt}$ is the velocity of the particles

$$\text{i.e.} \frac{\delta f}{\delta t} + \frac{\delta j}{\delta r} = 0, \tag{1}$$

where $j = f\dot{r}$ is the flux of particles

Now, the growth rate of any droplet is proportional to its size, r, and the concentration, $C(t)$, of the substance of the droplet in the medium (which is a constant according to assumption 3) with respect to its equilibrium value, C_{eq} :

$$\text{i.e.} \frac{dV}{dt} \propto r\left(C(t) - C_{eq}(r)\right),$$

$$\text{or } \frac{4\pi}{3}\frac{d}{dt}r^3(t) = 4\pi D\, r\left(C(t) - C_{eq}(r)\right), \tag{2}$$

where D is the molecular diffusivity of the disperse phase in the medium.

$$\text{or } r^2\frac{dr}{dt} = Dr\left(C(t) - C_{eq}(r)\right); \text{i.e.} \dot{r} = \frac{D}{r}\left(C(t) - C_{eq}(r)\right).$$

Now, we introduce a dimensionless quantity θ, which is a measure of the relative concentration by $\theta(t) = \left(\frac{C(t) - C_{eq}(\infty)}{C_{eq}(\infty)}\right)$; now from the Kelvin equation, we know that $C_{eq}(r) = C_{eq}(\infty)\left(1 + \frac{\alpha}{r}\right)$.

Therefore, $\dot{r} = \frac{D}{r}\left(C(t) - C_{eq}(\infty)\left(1 + \frac{\alpha}{r}\right)\right)$.

Rearranging the equation we get

$$\dot{r} = \frac{DC_{eq}(\infty)}{r}\left(\left(\frac{C(t) - C_{eq}(\infty)}{C_{eq}(\infty)}\right) - \frac{\alpha}{r}\right),$$

$$\text{or } \dot{r} = \frac{DC_{eq}(\infty)}{r}\left(\theta(t) - \frac{\alpha}{r}\right). \tag{3}$$

This is the expression of velocity of particle in size space. The value of particle radius at which the growth rate at any instant of time is zero ($\dot{r} = 0$) is called the instantaneous critical radius (r_c); then, it is obvious that $r_c = \alpha/\theta(t)$, and for growth rate equation, we can write $\theta(t) - \frac{\alpha}{r} > 0$ or $\frac{\alpha}{r} < \theta(t)$ or $r > \frac{\alpha}{\theta(t)}$, i.e. $r > r_c$, which means that at any instance of time particle with radius greater than the critical radius will exhibit an increase in size where the particle having radius smaller than the critical one will become smaller (decay). In this notation the growth rate equation (Eq. (2)) can be rewritten as follows:

$$\frac{4\pi}{3}\frac{d}{dt}r^3(t) = 4\pi D\, C_{eq}(\infty)\alpha\left(\frac{r}{r_c} - 1\right),$$

$$\text{or } \dot{r} = \frac{DC_{eq}(\infty)}{r^2}\left(\frac{r}{r_c} - 1\right).$$

The final element of LSW theory is the mass conservation.

In a closed system, the concentration of a substance in the medium and the drop size distribution function is interrelated by

$$\frac{d}{dt}\left[C(t) + \frac{4\pi}{3}\int_0^\infty r^3 f(r,t)dr\right] = 0,$$

$$\text{or } \theta(t) + \frac{4\pi}{3C_{eq}(\infty)}\int_0^\infty r^3 f(r,t)dr = \theta(0). \tag{4}$$

The three integro-differential Eqs. (1), (3) and (4) can be solved analytically. It was shown that this system should have an asymptotic solution which is independent of initial conditions. Rather than solving the problem for all times, LSW found an asymptotic solution valid as $t \to \infty$. Using this approach, the following predictions were made for two-phase mixture undergoing Ostwald ripening in the long-time limit [20]:

1. Time evolution of average droplet radius: $\overline{R}(t) = \left(\overline{R}^3(0) + 4t/9\right)^{1/3}$

2. Time evolution of the number of droplets per unit volume: $N(t) = \psi\left(\overline{R}^3(0) + 4t/9\right)^{-1}$

3. Time invariant droplet size distribution function: $f\left(\overline{R},t\right) = \frac{g(u)}{\overline{R}^4}$

where depending on the approach, $\overline{R}$ is either the critical radius (r_c) or the maximum drop size; $\overline{R}(0)$ is the average radius at the onset of coarsening; $\psi = \theta/\alpha\int_0^{3/2} u^3 g(u)du$, θ is the dimensionless concentration, $\theta = (C - C(\infty))/C(\infty)$; and u is the normalized radius, $u = R\overline{R}$, $\alpha = 4\pi/(3V_m C(\infty))$.

The *salient results of LSW theory* are summarized as follows:

1. In the stationary region, the nature of the size distribution function is time invariant.

2. The cube of the number-averaged particle size (r_N) varies linearly with time:

$$\omega = \frac{d}{dr}\left({r_N}^3\right) = \left(\frac{8\gamma DC(\infty)V_m}{9RT}\right) = \frac{4}{9}\alpha DC^\infty,$$

where r_c is the radius of droplet at steady state, D is the diffusion coefficient of the dispersed phase in the continuous phase and α is the characteristic length scale ($\alpha = 2\gamma V_m/RT$). Droplet with radius $r < r_c$ will disappear, while droplet with $r > r_c$ will grow.

The comparison between theoretical and experimental Ostwald ripening rates, however, evoked a significant discrepancy in the literature [18], where the latter was found to be several times higher than the former. It has been found that the linear relation of ${r_N}^3$ with time in the asymptotic limit does not always signify the Ostwald ripening; as a second mechanism, "Brownian-induced coalescence" (particularly if the drop surface coverage is not sufficient to hinder the coalescence) may also be operative, which has the same dependency of the rate over time. The LSW theory assumed that the droplets are fixed in space and the molecular diffusion is the only mechanism of mass transfer. However, for the case of droplets undergoing Brownian motion, one must take into account both the contribution of molecular and convection diffusion as predicted by Peclet number ($P_e = rv/\mathrm{D}$). The velocity v of a droplet of mass M is approximately given by $v = \left(\frac{3kT}{M}\right)^{1/2}$.

4.2 Adjustment in Ostwald ripening rate

The rate of ripening, according to the LSW model, is directly proportional to the solubility of the oil in the aqueous medium. The presence of amphiphiles (surfactants or co-surfactants) can significantly enhance the oil solubility by allowing them to enter into their hydrophobic core. By replacing the bulk solubility of oil, $C(\infty)$ by the concentration of oil solubilized by the micelles, and using the micellar diffusion coefficient instead of the molecular diffusion, one would get the Ostwald ripening rate in the presence of micelle. According to the *extended LSW theory*, the Ostwald ripening rate of nanoemulsion containing a water-insoluble low molecular weight coemulsifier (amphiphile) can be predicted by the following equation [24], $\omega = \left(\frac{64\gamma D_{co} C_{co}(\infty) V_m}{9RT}\right)\phi_{co}$, where D_{co}, $C_{co}(\infty)$ and ϕ_{co} are the molecular diffusivity, bulk solubility (in water) and volume fraction of the coemulsifier in the oil droplet, respectively. Similarly, as the interfacial tension (γ) is incorporated in the ripening rate equation in the capillary length parameter ($\alpha = 2\gamma V_m/\mathrm{RT}$), lowering of the interfacial tension by the amphiphiles will lead to smaller capillary length and therefore lower solubility of the oil at the droplet boundary. This will cause lowering of the ripening rate at least for the case of strongly adsorbed amphiphiles. In the absence of micelle, oil molecule transports into and through the aqueous phase separating the droplets. However, in the presence of micelle, it is proposed that mass transfer is still involved through the continuous phase, but the micelles increase the water solubility of the oil, therefore effectively increasing the transport rate [24–26]. Kabalnov [19] has proposed three possible mechanisms to explain the observed effects of micelles on Ostwald ripening; nevertheless, whether the oil molecules are taken up by the micelles directly from the aqueous medium or by fusion/fission of a micelle with a droplet surface is still unclear. Another possible way to slow down the Ostwald ripening rate was proposed by Higuchi and Misra [27] through the incorporation of a second disperse phase (oil) with a much lower continuous phase solubility. Initially, the concentration of the second insoluble phase is equal in all droplets. As the mixed oil nanoemulsion undergoes Ostwald ripening, the more soluble component diffuses from the smaller to larger droplet at a much faster rate than the less soluble one. Thus, time will come when the larger droplets become enriched with the soluble oil and the chemical potential of that component is equal in all the droplets. Thus, there is no driving force for further transfer of the oil to take place, and the state is referred to as pseudo-steady of ripening. As theoretically investigated by Kabalnov et al. in the case of medium-soluble second component, the Ostwald ripening rate of the mixed oil nanoemulsion can be approximated as [25]

$$\omega_{mix} = \left(\frac{\phi_1}{\omega_1} + \frac{\phi_2}{\omega_2}\right)^{-1},$$

where ϕ_1 and ϕ_2 and the ω_1 and ω_2 are the volume fraction and ripening rate of the medium-soluble and medium-insoluble component, respectively. In his treatment [25], Kabalnov came up with three stability regime depending on the mole fraction of the less soluble second component (X_2) in the mixture. For high amount of insoluble oil ($X_2 > 0.4$), entropy of mixing dominates, $\Delta\mu < 0$, and the nanoemulsion is thermodynamically stable. For low mole fraction ($X_2 < 0.17$), Laplace pressure between the different size droplets dominates, and the nanoemulsion is thermodynamically unstable. At intermediate mole fraction ($0.17 < X_2 < 0.4$), a nanoemulsion is said to be in metastable state. Initially, Laplace pressure will dominate and the system undergoes Ostwald ripening; however, it also has a kinetic energy barrier that prevents further ripening to occur.

The validity of the LSW theory was tested by Kabalnov et al. [18, 19]. The influence of the alkyl chain length of the hydrocarbon on the Ostwald ripening rate of nanoemulsion was systematically investigated from alkyl chain length C9–C16. Increasing the alkyl chain length of the hydrocarbon used for the emulsion results in the decrease in the oil solubility. According to the LSW theory, this reduction in solubility should result in a decrease in the Ostwald ripening rate which was confirmed by Kabalnov et al.

5. Stability against coalescence in nano-dispersion

Though the droplets coalescence rate is sufficiently get reduced for the nanoemulsion due to their smaller size, however, as a consequence of Ostwald ripening, it may results in creaming and ultimately leads to phase separation of the nanoemulsion (**Figure 6a**). The importance of droplet deformation, surfactant transfer and interfacial rheology for the stability of emulsion is well reported in the literature [28–30]. In the presence of thermal noise, the droplets are allowed to explore space by Brownian motion, and the resulting collisions cause the mean droplet radius R to increase. This process is called *coalescence*. Stabilization against

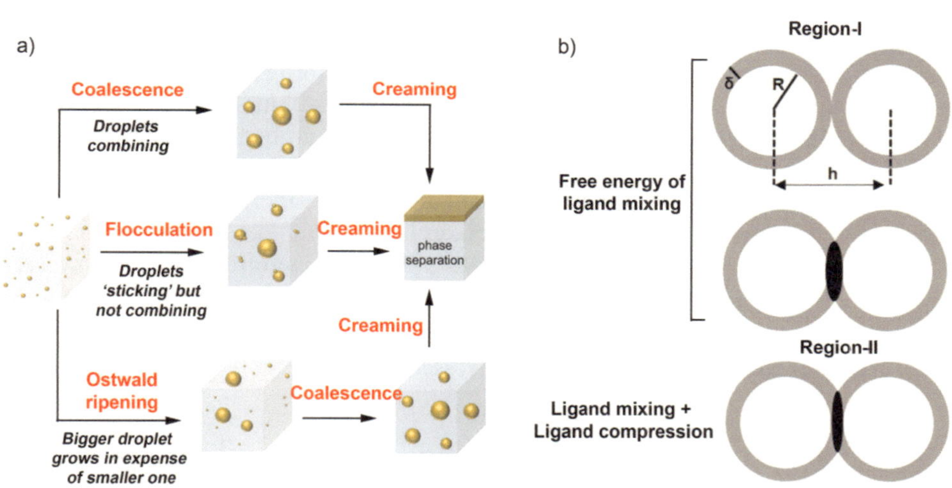

Figure 6.
(a) Schematic elaboration of various destabilization mechanisms for nanoemulsion and (b) cartoon of drop-drop interaction under two different regions where ligand shell overlaps each other.

coalescence can be effectively achieved, by creating a sufficient repulsive energy barrier between the droplets. This can be done by two ways, such as for O/W nanoemulsion the *electrostatic* stabilization (due to creation of double layer) by adding ionic surfactants. The formation of an electrical double layer (EDL, thickness δ) barrier is well established in the literature of colloidal stability according to the Derjaguin, Landau, Verwey and Overbeek (DLVO) theory [31]. As a result of this EDL formation when droplets approach a distance h, smaller than twice the double layer extension, strong repulsion occurs against their aggregation; hence, flocculation is prevented. A second and more effective mechanism using non-ionic surfactants or polymer (referred as surfactant) for W/O nanoemulsion is the *steric* stabilization (due to the presence of adsorbed polymer layer). However, no comparable definitive theory exists till date for the so-called steric stabilization. To date there have been several attempts made to develop the quantitative theory of steric stabilization with a notable success of *Fischer's solvency theory* [32] which exploits the Flory-Huggins theory [33] to predict the repulsive potential energy between two large flat plates. The total free energy of interaction obtained by Hesselink et al. for two flat plates coated by steric layer is given by [33]

$$\frac{U_{total}}{kT} = (2\pi/9)^{\frac{3}{2}}\gamma^2(\alpha^2 - 1)r_{rms}M(d) + 2\gamma V(d),$$

where γ is the number of ligands per unit area, α is the expansion factor, r_{rms} is the root mean square end-to-end distance of the chain in free solution and $M(d)$ and $V(d)$ are the distance-dependent mixing and elastic function that could be evaluated from segment density distribution function. However, the reason for the failure of Fischer's solvency theory appears to reside in the use of segment density distribution functions relevant to the isolated polymer according to Hesselink calculation [34] which is found unlikely to be applicable here. Smitham, Evans and Napper first pointed out the correction and proposed a simple analytical model to adopt for the segment density distribution function which is a constant segment density step function [35]. Their theory is able to account for many of the qualitative and quantitative features of steric stabilization observed to date. According to this model, the steric repulsion appears from two main origins: the *first one* (**Figure 6b**) is the unfavorable mixing of the surfactant chains which depends on their density at the interfacial region, on thickness of the interfacial layer (δ) and on the Flory-Huggin parameter, χ. The *second one* (**Figure 6b**) is the reduction in configuration entropy due to elastic stress of the chains which occur when inter-droplet distance becomes lower than δ. This method of stabilization is more effective than electrostatic stabilization in two ways: firstly, the repulsion is still maintained at moderate electrolyte concentration, and, secondly, the repulsion can be maintained at high temperature (provided ligand has solubility at that temperature).

The unfavorable mixing of the surfactant chains (considering the chain-solvent interaction predominates over the chain-chain interaction, like it is in good solvent condition) occurs when the ligand chain starts to interpenetrate within a distance, $1 + \overline{L} < h < 1 + 2\overline{L}$, where $\overline{L}$ is the rescaled length of ligand chain $\left(\overline{L} = \frac{\delta}{d}\right)$, δ is the contour length and $d(=2R)$ is the diameter of the sphere. The free energy of mixing in terms of rescaled parameter can be expressed as [35]

$$\frac{U_{mix}}{kT} = \frac{\pi d^3}{2V_{sol}^M}\varphi_{lig}^2\left(\frac{1}{2} - \chi\right)\left(h - (1 + 2\overline{L})\right)^2; \text{when, } 1 + \overline{L} < h < 1 + 2\overline{L},$$

where V_{sol}^M, φ_{lig}, χ and h are the molar volume of the solvent, the average volume fraction of the ligand segments in the overlapping region, the Flory-Huggin

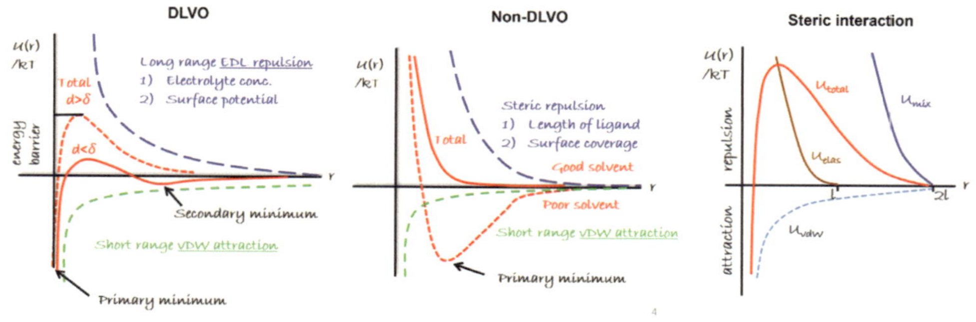

Figure 7.
Pictorial representation of distance dependence of pair potential energy between two spheres according to the literature.

interaction parameter and the center-to-center distance between any two approaching spheres, respectively. When $\chi < 0.5$, G_{mix} is positive and the interaction is repulsive. When $\chi > 0.5$, G_{mix} is negative and the interaction is attractive. When $\chi = 0.5$, G_{mix} is zero, which is referred to as the θ–condition. The Flory-Huggin parameter is calculated by $\frac{V_m}{RT}(\delta_s - \delta_m)^2 + 0.34$, where δ_s and δ_m are the Hildebrand solubility parameters [36] of the surfactant and solvent, respectively. Solubility parameter δ_2 of any component is related to its heat of vaporization ΔH with molar volume by $\delta_2 = \frac{\Delta H - RT}{V_m}$.

Continuing on the droplet-droplet approach within a distance $h < 1 + \overline{L}$, a second regime appears where chains undergo significant interpenetration and compression. The free energy of mixing in this regime can be expressed as [37]

$$\frac{U_{mix}}{kT} = \frac{\pi d^3}{2V_{sol}^M}\varphi_{lig}^2\left(\frac{1}{2} - \chi\right)\left[3\ln\left(\frac{\overline{L}}{h-1}\right) + 2\left(\frac{h-1}{\overline{L}}\right) - \frac{3}{2}\right], \text{ when } h < 1 + \overline{L},$$

while the elastic interaction U_{el} between the surfactant tails resulting from the loss in configurational entropy in this region is given by the expression [35]

$$\frac{U_{el}}{kT} = \pi\gamma d^2\left[(h-1)\left(\ln\frac{h-1}{\overline{L}} - 1\right) + \overline{L}\right]; \text{ when, } h < 1 + \overline{L},$$

where γ is the number of ligands per unit area of the sphere. Therefore, U_{el} is always repulsive (**Figure 7**). A pictorial representation of pair potential vs. interparticle distance was shown in **Figure 7** elaborating various cases of DLVO and non-DLVO interactions.

Therefore, the *criteria for effective steric stabilization* are the following: (1) the particle should be completely covered by the polymer, (2) the polymer should be strongly adsorbed to the particle surface, (3) the stabilizing chain should be highly soluble in the medium and is strongly solvated by its molecules and (4) δ should be sufficiently large to prevent weak flocculation which occurs when the thickness of adsorbed layer becomes small (<5 nm).

6. Conclusion

This chapter summarizes the most important aspects of nanoemulsion including the composition, structure and physical properties and also provides the glimpses of trademarks which distinguished it from the conventional microemulsion. Droplets having dimension in the order of few tens of nanometers are unstable due to the high difference in vapor pressure outside and inside the curved interface. The presence of polydispersity in the droplet size leads to the difference in their chemical potential (and hence solubility), which acts as a fuel for the transfer of mass between two droplets. However, with the help of a proper stabilizing layer of polymer (favorable volume to [head group area × length] ratio) or surfactant, we can achieve the necessary repulsive barrier (electrostatic, steric or electro-steric) against their coalescence. Concurrently though, it also raises the question of whether the additive (surfactant or polymer) can affect the rate of Ostwald ripening during the mass transfer process. With the proper knowledge of the surfactant properties (HLB and other surfactant parameters), one can effectively fabricate a stable nano-dispersion for a particular combination of dispersed and continuous phase.

7. Outlook

In order to leverage the knowledge of structure-property-function relationship for nanoemulsion, a continuous effort in research and development is required in this direction. The recent advances in nanoemulsion fabrication with reduced polydispersity in droplet size distribution are likely to provide new avenue to the researcher for achieving their desired functionality. Indeed, we are not so far to fabricate smart nanoemulsion (decorated, functionalized and internally structured) which can act as a vaccine or drug delivery vehicle to provide enhanced therapeutic response to biological system. Molecular design of amphiphiles can lead to new stable nanoemulsion topologies such as double nanoemulsion. Aggregate/gel can be formed by attractive nanoemulsion droplets by introducing strong opposite screening layer, or elastic verification in the system of repulsive nanoemulsion can be made through high-flow emulsification that cause extreme droplet rupturing.

Overall, nanoemulsions are very promising and flexible soft-matter systems, and they offer outstanding potential for new advances in basic science, customized colloidal design and high-value applications.

Acknowledgements

This material is based upon the work supported by the Science and Engineering Research Board (SERB), Department of Science and Technology, India. The author would like to thank Dr. B.L.V. Prasad.

DOI: http://dx.doi.org/10.5772/intechopen.84201

Author details

Kaustav Bhattacharjee
CSIR-National Chemical Laboratory, Pune, India

*Address all correspondence to: kaustavbhattacharjee@gmail.com

References

[1] Emulsions: Formation, Stability, Industrial Applications, by Tharwat F. Tadros, 2016. Walter de Gruyter GmbH, Berlin/Boston. e-ISBN (PDF) 978-3-11-045224-2

[2] Emulsion Science: Basic Principles An Overview, by Bibette J, Leal-Calderon F, Schmitt V, Poulin P. Springer-Verlag Berlin, Heidelberg, New York. ISSN electronic edition: 1615-0430

[3] McClements DJ. Nanoemulsions versus microemulsions: Terminology, differences, and similarities. Soft Matter. 2012;**8**:1719-11729. DOI: 10.1039/C2SM06903B

[4] Anton N, Vandamme TF. Nano-emulsions and micro-emulsions: Clarifications of the critical differences. Pharmaceutical Research. 2011;**28**: 978-985. DOI: 10.1007/s11095-010-0309-1

[5] Gupta A, Eral HB, Hatton TA, Doyle PS. Soft Matter. 2016;**12**:2826-2841. DOI: 10.1039/C5SM02958A

[6] Lautrup B. Physics of Continuous Matter. 2nd ed. Boca Raton: CRC press; 2011. DOI: 10.1080/00107514.2012.756936

[7] Tadros TF. Applied Surfactant: Principles and Application. Weinheim: Wiley; 2005. DOI: 10.1002/3527604812

[8] Tadros TF. Nanodispersions. Berlin/Boston: de Gruyter GmbH; 2016

[9] Wennerstrom H, Balogh J, Olsson U. Interfacial tensions in microemulsions. Colloids and Surfaces A: Physicochemical and Engineering Aspects. 2006;**291**: 69-77. DOI: 10.1016/j.colsurfa.2006.09.027

[10] Evilevitch A, Jonsson BT, Olsson U, Wennerstrom H. Molecular transport in a nonequilibrium droplet microemulsion system. Langmuir. 2001;**17**:6893-6904. DOI: 10.1021/la010899d

[11] Evilevitch A, Olsson U, Jonsson B, Wennerstrom H. Kinetics of oil solubilization in microemulsion droplets. Mechanism of oil transport. Langmuir. 2000;**16**:8755-8762. DOI: 10.1021/la000511z

[12] Hunter RJ. Foundations of Colloid Science. 2nd ed. Vol. II. London: Clarendon Press; 1989

[13] Israelachvili J. Intermolecular and Surface Forces. 2nd ed. Massachusetts: Academic Press; 1995

[14] Griffin WC. Classification of surface-active agents by HLB. Journal of the Society of Cosmetic Chemists. 1949; **1**:311 97

[15] Davies JT, Rideal EK. Interfacial Phenomena. 2nd ed. Massachusetts: Academic Press; 1963

[16] Beerbower A, Hill MW. Application of cohesive energy ratio (CER) concept to anionic surfactants. American Cosmetics and Perfumery. 1972;**87**: 85-94

[17] Hyde S, Andersson S, Larsson K, Blum Z, Landh T, Lidin S, et al. The Language of Shapes: The Role of Curvature in Condensed Matter: Physics, Chemistry and Biology. 1st ed. Amsterdam: Elsevier; 1997

[18] Kabalanov AS, Makarov KN, Pertsov AV, Shchukin ED. Ostwald ripening in emulsions: 2. Ostwald ripening in hydrocarbon emulsions: Experimental verification of equation for absolute rates. Journal of Colloid and

Interface Science. 1990;**138**:98-104. DOI: 10.1016/0021-9797(90)90184-P

[19] Kabalanov AS. Ostwald ripening and related phenomena. Journal of Dispersion Science and Technology. 2001;**22**:1-12. DOI: 10.1081/DIS-100102675

[20] Voorhees P. The theory of Ostwald ripening. Journal of Statistical Physics. 1985;**38**:231-252. DOI: 10.1007/BF01017860

[21] Kabalnov AS, Shchukin ED. Ostwald ripening theory: Applications to fluorocarbon emulsion stability. Advances in Colloid and Interface Science. 1992;**38**:69-97. DOI: 10.1016/0001-8686(92)80043-W

[22] Lifshitz IM, Slyozov VV. The kinetics of precipitation from supersaturated solid solutions. Journal of Physics and Chemistry of Solids. 1961;**19**:35-50. DOI: 10.1016/0022-3697(61)90054-3. (Translated by Lowde RD)

[23] Wagner C. Theory of the aging of precipitation by redissolution (Ostwald ripening). Zeitschrift für Elektrochemie. 1961;**65**:581-591. DOI: 10.1002/bbpc.19610650704

[24] Kabalnov AS. Can micelles mediate a mass transfer between oil droplets? Langmuir. 1994;**10**:680-684. DOI: 10.1021/la00015a015

[25] Kabalnov AS, Pertsov AV, Shchukin ED. Ostwald ripening in two-component disperse phase systems: Application to emulsion stability. Colloids and Surfaces. 1987;**24**:19-32. DOI: 10.1016/0166-6622(87)80258-5

[26] Smet YD, Deriemaeker L, Finsy R. Ostwald ripening of alkane emulsions in the presence of surfactant micelles. Langmuir. 1999;**15**:6745-6754. DOI: 10.1021/la9901736

[27] Higuchi WI, Misra JJ. Physical degradation of emulsions via the molecular diffusion route and the possible prevention thereof. Journal of Pharmaceutical Sciences. 1962;**51**: 459-466. DOI: 10.1002/jps.2600510514

[28] Ivanov IB, Danov KD, Kralchevsky PA. Physical degradation of emulsions via the molecular diffusion route and the possible prevention thereof. Colloids and Surfaces A: Physicochemical and Engineering Aspects. 1999;**152**:161-182. DOI: 10.1016/S0927-7757(98)00620-7

[29] Petsev DN, Denkov ND, Kralchevsky PA. Flocculation of deformable emulsion droplets. Journal of Colloid and Interface Science. 1995; **176**:201-213. DOI: doi.org/10.1006/jcis.1995.0023

[30] Denkov ND, Petsev DN, Danov KD. Flocculation of deformable emulsion droplets. I. Droplet shape and line tension effects. Journal of Colloid and Interface Science. 1995;**176**:189-200. DOI: 10.1006/jcis.1995.0022

[31] Verwey EJ, Overbeek JTG. Theory of the Stability of Lyophobic Colloids. 1st ed. Amsterdam: Elsevier; 1948

[32] Fischer EW. Electron microscopic studies on the stability of suspensions in macromolecular solutions. Kolloid-Zeitschrift. 1958;**160**:120-141. DOI: 10.1007/BF01512392

[33] Flory J. Principles of Polymer Chemistry. Ithaca: Cornell University Press; 1953

[34] Hesselink FT, Vrij A, Overbeek JTG. Theory of the stabilization of dispersions by adsorbed macromolecules. II. Interaction between two flat particles.

Journal of Physical Chemistry. 1971;**75**: 2094-2103. DOI: 10.1021/j100683a005

[35] Smitham JB, Evans R, Napper DH. Analytical theories of the steric stabilization of colloidal dispersions. Journal of the Chemical Society, Faraday Transactions. 1975;**1, 71**: 285-297. DOI: 10.1039/F19757100285

[36] Hildebrand JH. Solubility of Non Electrolytes. 2nd ed. New York: Reinhold; 1936

[37] Evans R, Smitham JB, Napper DH. Theoretical prediction of the elastic contribution to steric stabilization. Colloid & Polymer Science. 1977;**255**: 161-167. DOI: 10.1007/BF01777275

Chapter 7

Synthesis, Properties, and Characterization of Field's Alloy Nanoparticles and Its Slurry

Chaoming Wang, Xinran Zhang, Wenbing Jia, Wei Wu and Louis Chow

Abstract

This chapter describes a facile one-step method developed for the synthesis of Field's alloy nanoparticles using a nanoemulsification technique and their dispersed them in a base fluid to make slurry. The composition, size, morphology, and thermal properties of as-prepared nanoparticles were characterized by XRF, TEM and, DSC, respectively. The slurry with Field's alloy nanoparticles exhibited good thermal properties and stability. Meanwhile, an experimental study was performed to investigate the jet impingement of HFE7100 fluid with nanosized metallic (Field's alloy) phase change materials (nano-PCM). Surface modification was used to stabilize the slurry of the nano-PCM in HFE7100 fluid and make the slurry stable for over 1 month. The Field's alloy nano-PCM absorbed heat during a phase change process from solid to liquid phase coupled with HFE7100 evaporation process. The effects of mass fraction of Field's alloy nano-PCM on the pressure drop and heat transfer performances of the slurry were investigated through a heat transfer loop test. Away from the critical heat flux, Field's alloy nano-PCM slurry provided a significant heat transfer enhancement due to the increase in the thermal capacity of the carrier fluid. Moreover, the nano-PCM slurries were able to maintain 97% of their heat removal capability after 5000 thermal cycles.

Keywords: Field's alloy, nanoparticles, nanoemulsification, slurry, HFE7100, jet impingement heat transfer

1. Introduction

With their wide industrial and civil applications, heat transfer fluids (HTFs) have been potentially used in lubrication, energy storage, heat exchange, electronic cooling, and so on [1–5]. However, for conventional HTFs (such as water, polyalphaolefin (PAO), fluorocarbons, and glycols), the main drawback of deficient heat transfer performance owing to their low thermal conductivities has limited their practical applications. For the purpose of improving their heat transfer properties, earlier research efforts have been carried out by dispersing those materials, which have high thermal conductivities, such as silver, copper, alumina, copper oxide, silicon carbide, and carbon nanotubes, into HTFs [6–12].

Until recently, adding nano-sized phase change materials (nano-PCMs) into base HTFs attracts considerable attentions. The most frequently used PCMs include inorganic PCMs, such as metal, alloy and salt hydrates, and organic PCMs, such as paraffins, polyethylene glycols, fatty acids, and esters [13–19]. Different types of nano-PCMs can be synthesized by using various synthetic methods [20–22]. By encapsulating or coating the nano-PCM with a suitable layer, the nano-PCMs can be dispersed in a base fluidic phase. However, nano-sized particles have a strong tendency to agglomerate and easily lead to precipitate in HTFs, which restricts their application as thermal energy storage media. Therefore, encapsulation or surface modification of nanoparticles to increase their dispersion ability in the carrier fluid is of primary importance. Various techniques such as interface polymerization [23] and coacervation [24] and emulsion polymerization [25, 26] were explored to make encapsulated nano-PCM. Meanwhile, some modified nanoparticles using certain way exhibit excellent dispersion stability in some HTFs [27–29]. As inexpensive and stable dielectric HTFs, PAO and HFE7100 are usually applied in cooling of avionic systems [30, 31].

Among those PCMs, the thermal conductivity of lots of low melting point metals, such as indium (In), bismuth (Bi), tin (Sn), and lead (Pb) and their eutectics, is at least two orders higher than that of inorganic PCMs. Meanwhile, their latent heat density and other thermal properties are comparable to inorganic PCMs, which makes low melting point metals or eutectic alloys highly attractive PCMs in practical applications. By taking advantages of nano-PCMs (such as their small size, large surface-to-volume ratio, good dispersion ability in base fluid, and large latent heat of fusion), these HTFs have some distinct merits such as high energy density thermal storage, large specific heat capacity, low flow drag, and enhanced thermal conductivity. Meanwhile, the fluids still keep the fluidic properties. All these aforementioned merits make these thermal fluids containing nano-PCM a promising HTF for electronic cooling equipment, thermal control, and those systems requiring high heat transfer rates [32–34]. The tight contact of nanoparticle and base fluid decreases the heat transfer resistance between nanoparticles and fluid, thus enabling fast exchange of heat transfer between phases [35]. Therefore, the slurry with nano-PCM could decrease the total pumping power in a heat transfer loop due to the increased heat capacity of the carrier fluid.

For a liquid, when the flow rate and thermal conductivity keep constant, the heat transfer capability is predominantly depending on its heat absorbing capacity [36]. Frequently, in high-flux heat removal case, dielectric fluids (such as HFE7100) are usually used to take the heat away through utilizing their latent heat of vaporization. As a low melting point alloy, Field's alloy is a eutectic alloy melt at approximately 62°C (144 F), in which composes with the weight percentage of 32.5% bismuth (Bi), 51% indium (In), and 16.5% tin (Sn). As a low melting point alloy, Field's alloy is selected due to its melting temperature a litter higher than the boiling point of HFE7100. Therefore, during liquid-vapor phase transition of HFE7100, the thermal fluid heat capacity can be increased significantly when the nano-PCM changes from the solid to liquid phase.

This chapter mainly summarized our works in recently few years [37, 38], which include: (1) synthesis and modification of Field's alloy nanoparticles; (2) characterization of as-prepared Field's alloy nanoparticles; (3) jet impingement heat transfer of Field's alloy nanoparticles-HFE7100 slurry.

2. Synthesis and modification of Field's alloy nanoparticles

Nanoemulsification method is one of the most facile techniques to prepare nano-sized Field's alloy particles. The illustrated nanoemulsification formation process is shown in **Figure 1**. Briefly, a certain amount of Field's alloy pellets was put into PAO

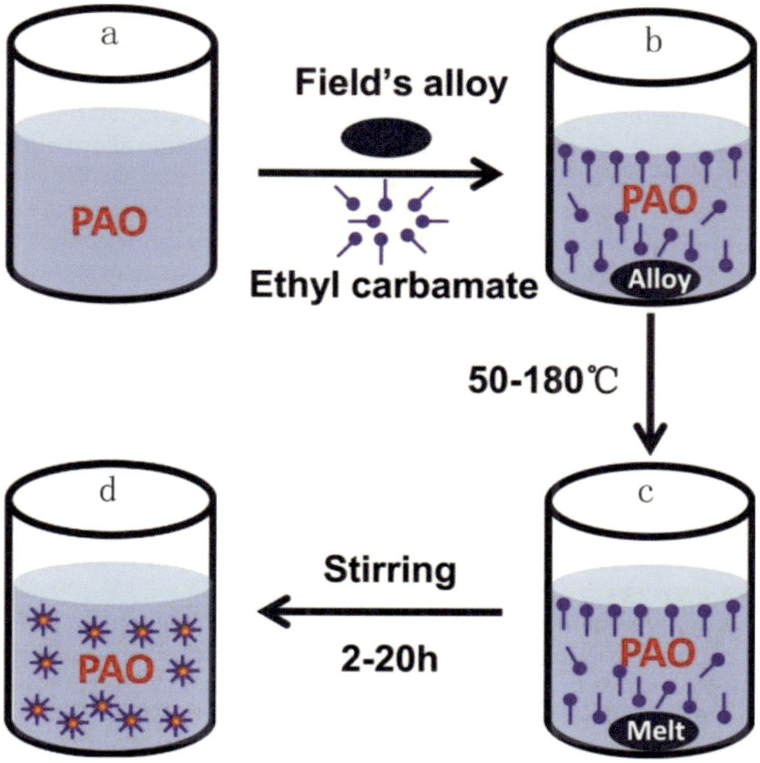

Figure 1.
Illustrated scheme of the synthesis of molten Field's alloy nanoparticles using nanoemulsification method. (a) PAO and molten Field's alloys are in the reaction vessel. These two liquids are immiscible and phase separate; (b) polymer surfactant (ethyl carbamate) is soluble in PAO; (c) the mixture is heated up to certain temperature and the bulk molten alloy formed; (d) the microscale emulsion is stirred and breaks into microscale droplets until nanoemulsion is formed.

oil with or without certain amount of ethyl carbamate as surfactant. After that, the mixture was heated and kept at certain temperature (50, 70, 100, 150, and 180°C) in the help of silicone oil thermal bath and stirring for specific time under certain temperature under nitrogen protection to make the native or ethyl carbamate modified low melting temperature Field's alloy nanoparticles. As the time increases, the white color of PAO oil became gray and dark gradually. When the reaction finished, the nanoparticles were gathered by centrifuge and washed with acetone at least three times and then dried at 45°C overnight for ready to use. The native or modified Field's alloy slurry was made by dispersing certain amount of native or ethyl carbamate modified Field's nanoparticles into the desired base fluid.

In order to make the as-prepared Field's alloy nanoparticles dispersed well in HFE7100 for long time, the nanoparticle surfaces were modified with a monolayer of 1H, 1H, 2H, 2H-perfluorooctyltriethoxysilane by mixing the nanoparticles in a silane solution of HFE7100.

3. Characterization of as-prepared Field's alloy nanoparticles

The size, morphologies, composition, and thermal properties of the synthesized Field's alloy nanoparticles were characterized by transmission electron microscopy (TEM), X-ray fluorescence spectrometry (XRF), and differential scanning calorimetry (DSC). The result and discussions were as the following.

3.1 Size and morphologies of Field's alloy nanoparticles

The size and morphologies of synthesized Field's alloy nanoparticles using nanoemulsion method were investigated by TEM. **Figure 2** showed the TEM images of the as-prepared Field's alloy particles after boiling the Field's alloy pellets in PAO at 180°C for 10 min (**Figure 2A**) and 2 h (**Figure 2B**), respectively. When the boiling time increases, the size of particles would decrease and became more and more spherical. After 10 min, the particles have irregular shapes and with the size range from 200 to 500 nm. As the time increases to 2 hours, the particles showed spherical shapes with the size of about 20 nm. **Figure 2C** showed the plot of reaction time versus the size of particles. After 2 hours, even though the reaction time increase (such as 5, 10, or even 20 h), the particles size did not have too much change, which was still close to 20 nm.

3.2 Composition and thermal properties of Field's alloy nanoparticles

The composition of bulk Field's alloy pellet and the synthesized Field's alloy nanoparticles was deduced with XRF spectrum, which was measured and collected from a mini-X system that uses a mini X-ray tube (Amptek, 40 kV, 100 μA) and a solid state X-ray spectrometer detector (Amptek 123, reflection mode). The whole setup was enclosed in a lead containing acrylic chamber with 1 mm of lead equivalent thickness to make sure no X-ray comes out. As shown in **Figure 3A**, no obvious difference is observed between the composition of the bulk Field's alloy materials and the as-prepared nanoparticles, in which $L_{\alpha1}$, $K_{\alpha1}$, and $K_{\alpha2}$ of indium at 3.29,

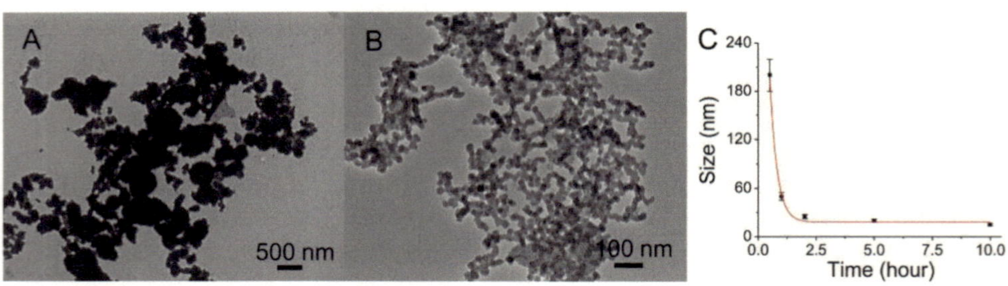

Figure 2.
TEM images of Field's alloy particles synthesized at 180°C for 10 min (A) and 2 h (B) using nanoemulsification method; Field's alloy particles size versus reaction temperature for 2 h (C).

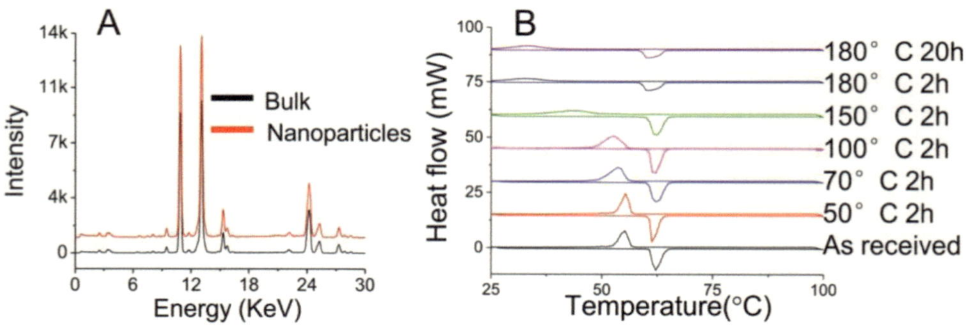

Figure 3.
X-ray fluorescence spectrum of bulk Field's alloy (black) and nanoparticles (red) (A); DSC curves of Field's alloy nanoparticles synthesized at different temperatures (50, 70, 100, 150, 180°C) for 2–20 h, respectively (B).

24.21, and 27.27 keV; $L_{\alpha 1}$ and $L_{\beta 1}$ of bismuth at 10.84 and 13.02 keV; and $K_{\alpha 1}$ of tin at 25.27 keV can be seen clearly. It indicated that no phase separation happens during the whole synthesis process.

DSC was used to measure the thermal properties of as-prepared nanoparticles. **Figure 3B** showed the DSC curves of bulk Field's alloy pellets and corresponding nanoparticles synthesized at different temperatures (50, 70, 100, 150, and 180°C) with other experimental conditions unchanged. For melting peaks of those samples, the peak position and shape were close and similar to the bulk material, where all of the samples are melt at about 62.5°C. However, for freezing peaks, it showed different characteristics for those nanoparticles synthesized at different temperatures. As the temperature increases, the freezing peak position would decrease and became more and more broaden. For example, the bulk Field's alloy was freezing at 55.2°C, and the nanoparticles synthesized at 50°C for 2 h, 70°C for 2 h, 100°C for 2 h, 150°C for 2 h, 180°C for 2 h, and 180°C for 20 h are at 55.2, 53.9, 52.5, 43.4, 32.6, and 32.6°C, respectively. Comparing the DSC curves of those samples under different temperatures, the reaction temperature-dependent freezing depressing was observed.

3.3 Thermal stability of Field's alloy nanoparticles and slurry

The thermal properties of Field's alloy slurry were investigated using a DSC. As shown in **Figure 4A**, the slurry can undergo melting-freezing phase transition during heating and cooling scanning processes. The downward endothermic peak at 62.5°C was belonging to the melting of the Field's alloy nanoparticles, while the upward

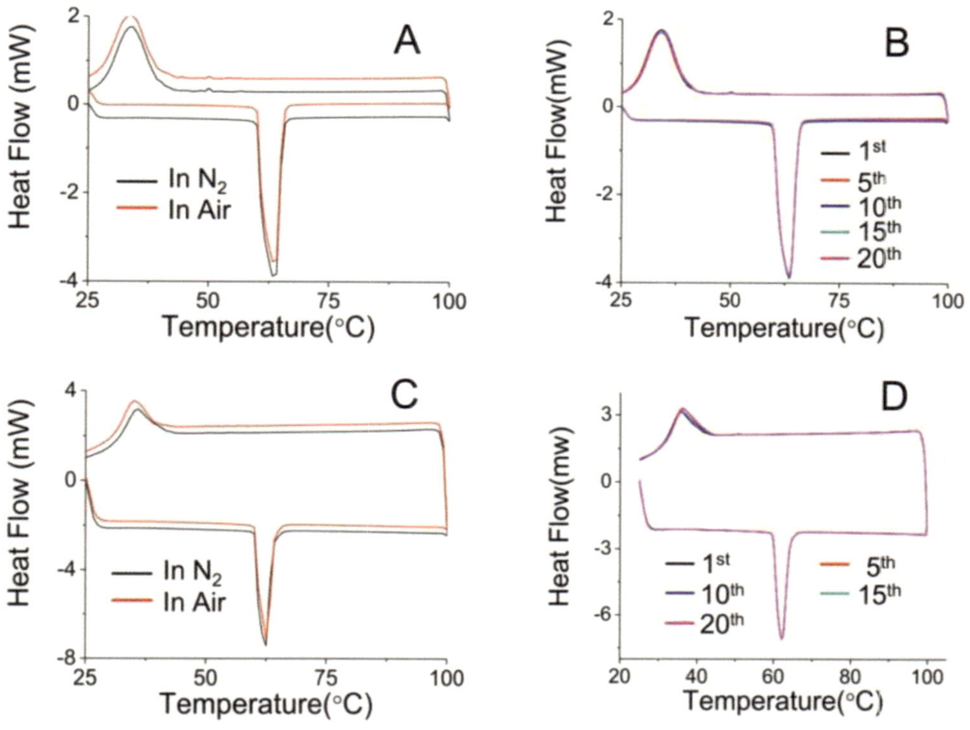

Figure 4.
Cyclic DSC curves of Field's alloy nanoparticles in ambient and nitrogen atmosphere, respectively (A); cyclic DSC curves of Field's nanoparticles after running in ambient condition for 1st, 5th, 10th, 15th, and 20th times, respectively (B); cyclic DSC curves of the slurry in ambient and nitrogen atmosphere (C); cyclic DSC curves of the slurry after running for 1st, 5th, 10th, 15th, and 20th times in ambient condition, respectively (D).

exothermic peak at 35.5°C was for their freezing. This observed melting-freezing temperature difference (about 27°C), called supercooling, had been well explained by the classical nucleation theory [39]. As shown in **Figure 4B**, no obvious changes of the slurry after running for 20 cycles, it suggested that slurry was very stable in ambient condition. It suggested that the nanoparticles are stable in ambient condition. As shown in **Figure 4C**, the nano-PCM slurry can undergo melting-freezing phase transition during heating and cooling scanning processes. The downward endothermic peak at 62.5°C was belonging to the melting of the Field's alloy nanoparticles, while the upward exothermic peak at 35.5°C was for their freezing. As shown in **Figure 4D**, no obvious changes of the nano-PCM slurry after running for 20 cycles, it suggested that nano-PCM slurry is very stable in the temperature range under 100°C.

4. Heat transfer loop test

In order to investigate the Field's alloy nanoparticle slurry compared to the pure liquid of HFE7100, a jet impingement heat transfer test was carried out. This part will represent some experimental data and discuss the effect of mass fraction of Field's alloy nanoparticle in slurry on pressure drop and heat transfer performance.

4.1 Experimental setup for heat transfer loop test

It is noted that all of the heat transfer loop tests were carried out under 1 atmospheric pressure. After installation of the heater, the test vessel was evacuated and filled with the working fluid, HFE7100. Additional degassing process was carried out by boiling the liquid pool for 2 hours to remove the dissolved noncondensibles. **Figure 5** illustrated the flow loop utilized to conduct the experiments. A variable

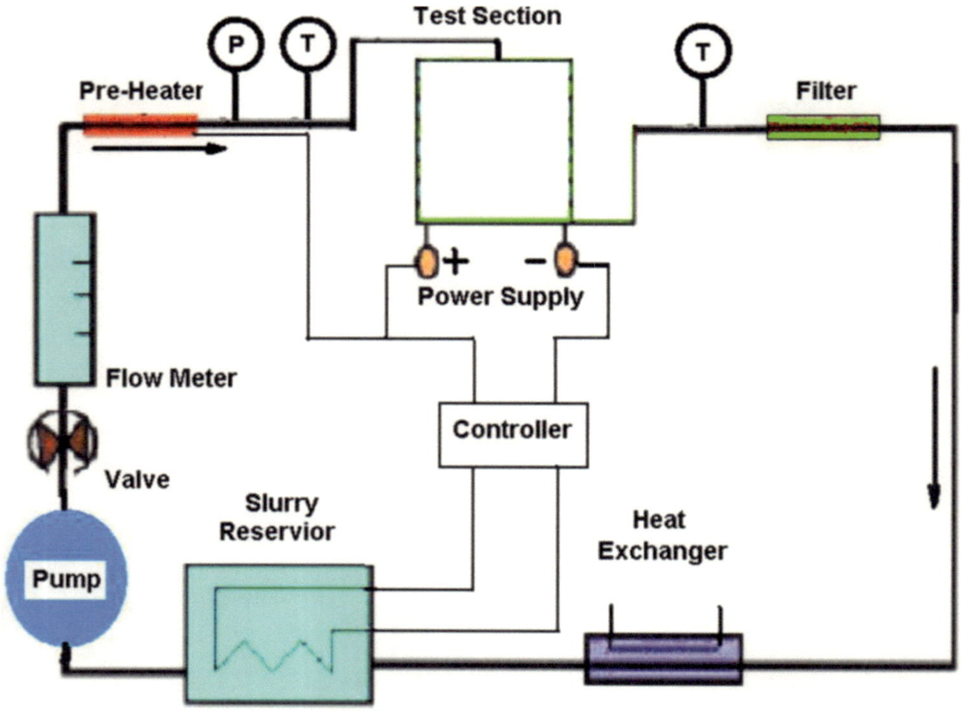

Figure 5.
Schematic diagram of the heat transfer loop test.

speed gear pump was used to supply coolant to the nozzle. A turbine flow meter was used to measure the volume flow rate. A vertically oriented nozzle was located directly above the heater surface. The distance between the nozzle exit and the test surface was fixed at 20 mm using a fine-threaded post-arrangement. The nozzle heater assembly was located in a chamber which also acts as the coolant reservoir.

4.2 Test chamber and heaters

Figure 6A illustrated the test chamber and **Figure 6B** showed our measured relationship between chamber pressure and saturation temperature of HFE7100.

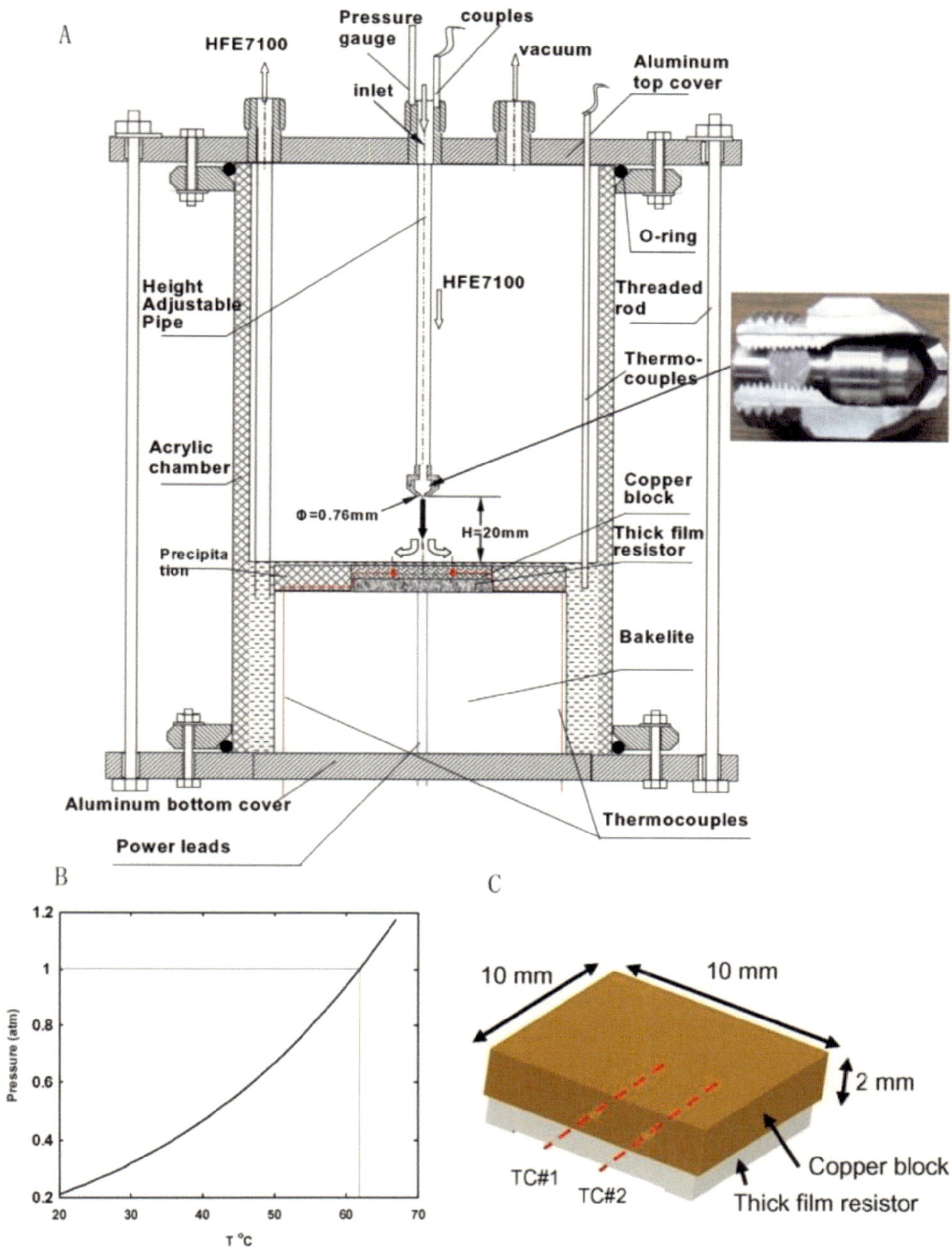

Figure 6.
Illustrated scheme of test chamber (A) temperature vs. pressure of HFE7100 (B) and heater (C).

The coolant within the reservoir was maintained at a constant temperature using a combination of an immersion heater and proper thermal insulation to ensure the test was under 1 atmosphere pressure. An acrylic cylinder with aluminum lids, which was 300 mm tall and 200 mm in diameter, was used as the test chamber. The test chamber contained approximately 500 ml of working slurry (pure HFE7100/ nano-PCM mixture). Heat was applied to the heater surface through a copper plate using a resistive heater controlled by a HP 6030A DC power supply system. The heater was built by soldering a 10×10×2-mm-copper block onto a matching size, 1-mm-thick resistive heater. **Figure 6C** showed the heater details. The resistive heater was made of a 5-ohm-thick film resistor with BeO substrate made by Barry Industries Inc. Soldering the thick film resistor to the copper plate minimizes the thermal contact resistance at the interface.

The temperature of slurry entering the test chamber was maintained at 56°C, 5°C below the liquid saturation temperature. Due to the large supercooling of nano-PCM particles (solidification temperature at 17.1°C), to ensure the PCM in solid phase, the slurry was chilled to 5°C by ice water. The nozzle inlet temperature (56°C) was controlled by an auxiliary heater powered by an AC regulator. The test heater surface was fastened at the bottom of the container using epoxy resin. The test vessel was insulated with a 25-mm-thick fiber glass blanket. A Bakelite layer (with a thermal conductivity less than 1 W/mK) underneath the heater was found to be a sufficient insulator. The uniformity of the Joule heating over the resistor surface was within 5%. The heat fluxes were also controlled at steady state for each set of testing. The jet impingement nozzle (a TG 0.7 full cone nozzle with the insert removed) was procured from Spraying Systems Co., and it had a passage with circular cross section, 0.76 mm in diameter. The coolant was allowed to enter the plenum of the nozzle where it flowed through a converging section as it existed to the chamber. Once the steady-state flow and temperature conditions were attained, the mean temperature of the heater and inlet and outlet temperatures of the slurry were calculated by the arithmetic mean of temperature readings. The heat transfer coefficient was then obtained by the following equation [40]:

$$h = Q/A(T_w - T_f) \tag{1}$$

where T_w is the average of the two surface temperatures extrapolated from the two embedded thermocouple readings. The pressure drop across the nozzle was measured at different mass fractions of slurry at a temperature of 20°C at the nozzle inlet (non-melting conditions).

4.3 Physical properties of Field's alloy nanoparticle slurry in HFE7100

For the silane monolayer modified Field's alloy nanoparticles, the optical transmittance of nanoparticle suspensions in HFE7100 fluid was monitored by a portable spectrometer, and the optical transmittance kept nearly the same over 1 month, which indicates the nanoparticle suspension is highly stable for more than a month after the 1H, 1H, 2H, 2H-perfluorooctyltriethoxysilane modification. The physical properties of nano-PCM slurry in HFE7100 with 10, 23, and 30% particle mass fraction, where the corresponding particle volume fractions of slurry are 2.3, 6.4, and 8.0%, and its components are presented in **Table** 1. The bulk viscosity of Field's alloy nanoparticle slurry is measured by using a calibrated Cannon-Fenske viscometer. Other parameters, such as density, thermal conductivity, and latent heat of Field's alloy nano-PCM slurry, are calculated according to the reference data and mixture equations [41]. The specific heat of nanoparticles for solid/liquid phases is derived from the superposition calculations involving HFE7100 and Field's alloy (solid/liquid).

Slurry and its components	Density (kg m^{-3})	Specific heat (J $kg^{-1}K^{-1}$)	Thermal conductivity (W $m^{-1}K^{-1}$)	Latent heat (kJ kg^{-1})	Viscosity (mPa s) at 293 K
Field's alloy[a] (solid)	7880	170.5	70.1	40.2	–
Field's alloy (liquid)	7880	170.5	34.5	–	–
HFE7100 (298 K)[b]	1500	1180	0.07	–	0.60
Slurry[c] 10%	1648	1079	0.075	4	0.69
23%	1883	948	0.084	9	0.75
30%	2036	877	0.091	11	0.79

[a]*http://www.Matweb.com [42].*
[b]*3M Data Book, HFE7100 for Heat Transfer, 2002 [43].*
[c]*Bulk physical properties of slurries are calculated from those of solid PCM particles.*

Table 1.
Physical properties of Field's alloy nano-PCM slurry and its components.

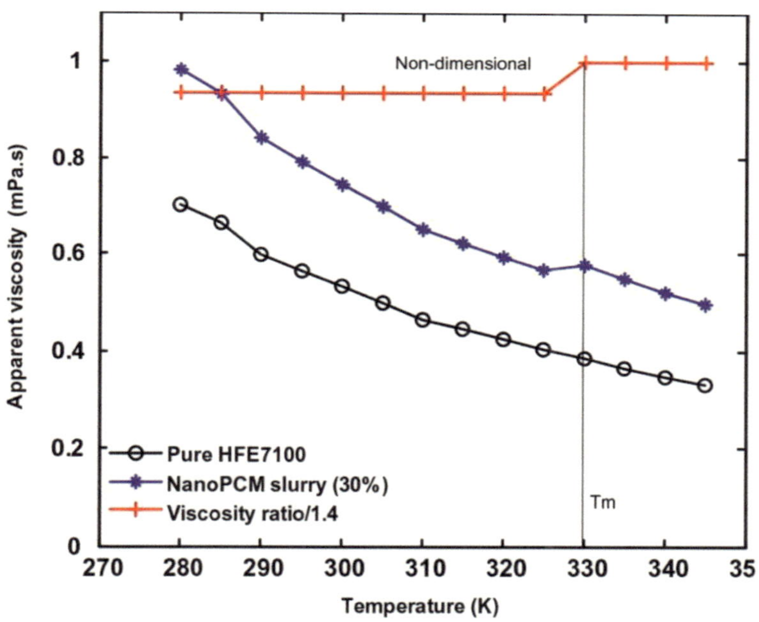

Figure 7.
Effect of temperature on apparent viscosity of pure HFE7100 and nano-PCM slurry.

A relation exists between the temperature of HFE7100 with Field's alloy nanoparticles slurry and its apparent viscosity. **Figure 7** shows the measured experimental data that exhibits the effect of temperature on apparent viscosity of pure HFE7100 and slurry with 30% (wt%) particle fraction. The results show that the ratio of the slurry viscosities keeps almost the same at 1.4 over the tested temperature range.

4.4 Pressure drop

An important parameter was pressure drop between the inlet and the outlet of microchannel heat exchanger. In ideal case, the pressure drop should be as small

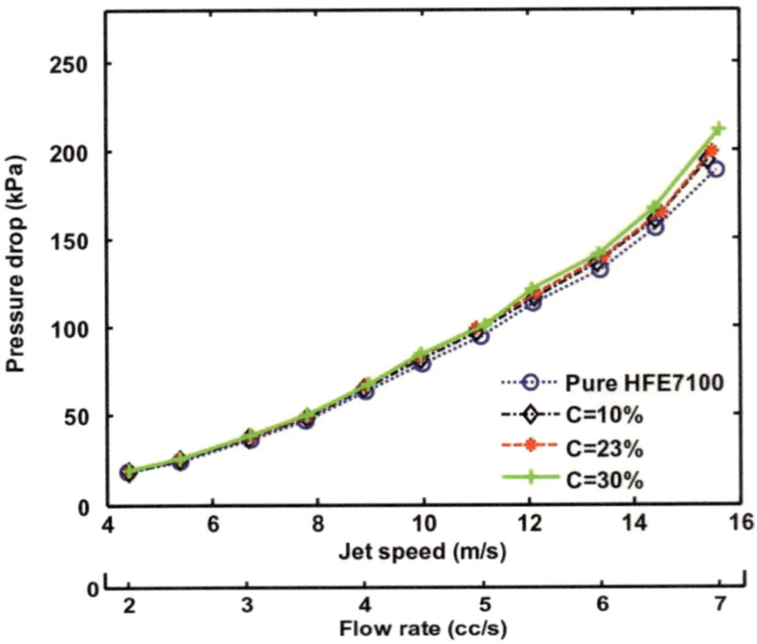

Figure 8.
Jet impingement pressure drop data with different Field's nanoparticle concentrations (nozzle: TG0.76, H = 20 mm); the inlet temperature is controlled at 20°C.

as possible to minimize pumping power. When the velocity of the jet was varied between 4.2 and 15.5 m/s, the pressure drop across the nozzle was measured at flow rates of 1.95–7.15 cc/s. As shown in **Figure 8**, when the nozzle inlet temperature of the liquid temperature was controlled at 20°C, the jet impingement pressure drop results were very close even with different particle mass fractions. It can be assigned to the comparable increases in Field's alloy nanoparticle slurry viscosity and density with increased nanoparticle mass fraction (see **Table** 1). The pressure drop was related to the nozzle Reynolds number as $\triangle P \propto Re_d^{-0.25}$ [44]. Since the nozzle Reynolds number did not change too much with increased particle mass fraction, the pressure drop should not vary significantly with particle mass concentration.

4.5 Heat transfer performance of Field's alloy nanoparticle slurry

Heat transfer performance of the nanoparticle slurries was evaluated by measuring their convective heat transfer coefficients in jet impingement configuration. When no solid-liquid and liquid-vapor phase change occurs in the slurry, the jet impingement heat transfer coefficient can be predicted by the Martin correlation [45]. **Figure 9** showed the heat transfer coefficient of jet impingement test for pure HFE7100 when the inlet nozzle temperature was kept at 20°C with a flow rate from 2.05 to 6.95 cc/s. When the flow rate was from 2.05 to 6.95 cc/s, the heat transfer coefficient was increased from 4.6 to 9.4 103 W/m^2 K. The difference between the experimental data and the Martin correlation was within 10% of the Martin correlation.

In order to compare the heat transfer performance between HFE7100 and the slurries with different particle mass fractions, the flow rate constant was set at 7.15 cc/s, and the temperature of the liquids at the nozzle inlet was fixed at 56°C. The heater surface temperature was varied between 56 and 75°C by controlling the heater power input. The comparison for the overall performance of pure HFE7100 (0%) and slurries with 10, 23, and 30% particle mass fraction, using heat flux as a measure, was provided in **Figure 10**. It can be seen that even before the melting temperature of

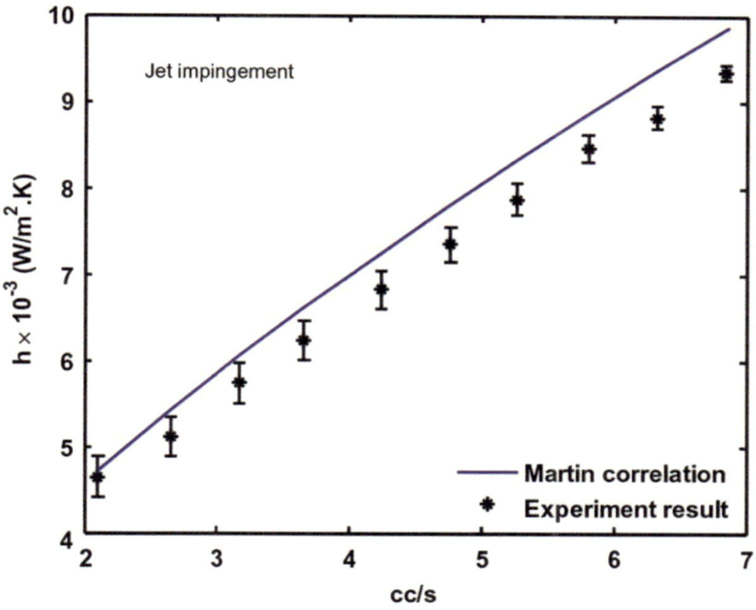

Figure 9.
Experimental and Martin correlation of heat transfer coefficients of pure HFE7100 when the inlet temperature is 20°C.

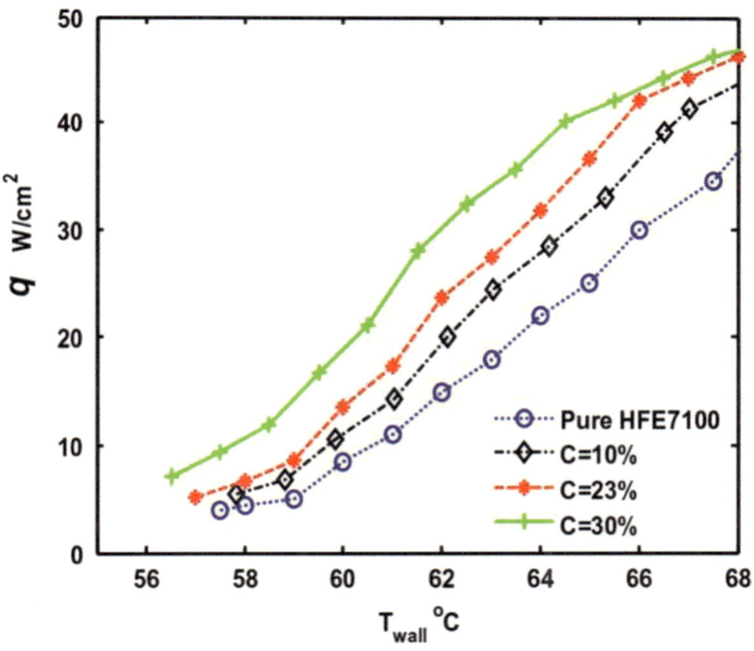

Figure 10.
HFE7100 jet impingement (56–68°C) heat flux at the flow rate of 7.15 cc/s; (note: the inlet T_{in} was controlled at 56°C, nozzle: TG0.76, H = 20 mm). Slurry with 30% particle mass fraction improved average heat transfer coefficient by 70% when compared to pure HFE7100 for jet impingement at the temperature range from 62 to 66°C.

Field's alloy and the boiling point of HFE7100 were reached, the slurries had a higher heat flux at surface temperature between 56 and 60°C. This can be attributed, in part to their higher thermal conductivity. At the heater surface of 62°C, the results clearly

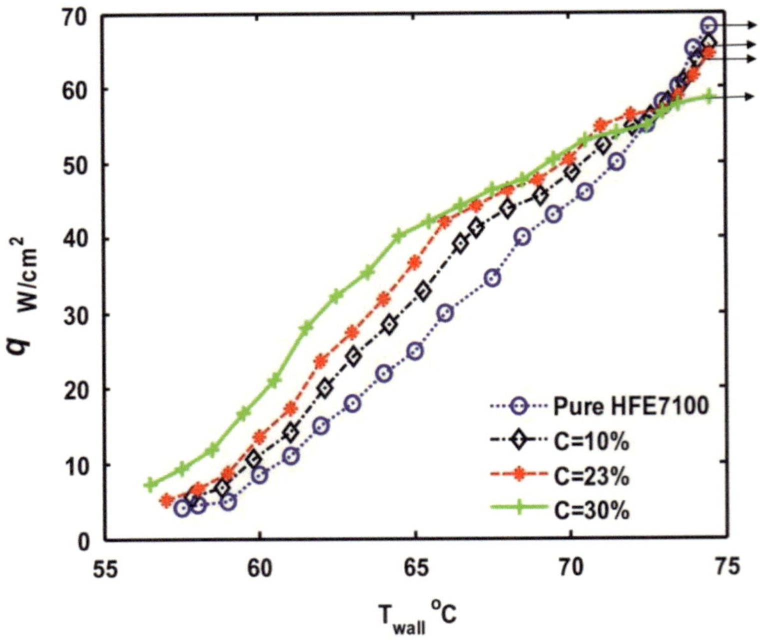

Figure 11.
Heater surface temperature at full range (56–75°C) at the flow rate of 7.15 cc/s; (note: the inlet T_{in} was controlled at 56°C, nozzle: TG0.76, H = 20 mm), black arrow was the critical heat flux.

indicated that the slurry with 30% particle mass fraction provides a much higher heat flux than pure HFE7100. The heat flux removal increased from 15 to 30 W/cm^2, a 100% improvement. At 64°C, the results showed that the slurry with 30% particle mass fraction increases heat flux from 22 to 38 W/cm^2, a 73% improvement. The results also show that at 66°C, the slurry with 30% particle mass fraction had a higher heat flux, 30 W/cm^2 vs. 44 W/cm^2, a 47% improvement. On the average, the slurry with 30% particle mass fraction provided a heat transfer enhancement of 70% when compared to pure HFE7100.

It was noted that the heat transfer enhancement was heat-flux dependent. It begun to decrease as the heat flux increased because of the increasingly higher temperature difference between the wall and the slurry. High heat flux can shift the slurry temperature out of the melting range (>62°C). **Figure 11** showed the heat removal results at a flow rate of 7.15 cc/s over a wider surface temperature range, between 56 and 75°C. It was interesting to note that, at surface temperature higher than 73°C, the critical heat flux decreased as the particle fraction increased from 10 to 30%. This could be explained by the increase in viscosity and latent heat of melting depletion. These combined effects could reduce the overall heat removal capability of nano-PCM slurry. At a surface temperature of 75°C and flow rate of 7.15 cc/s, the heat fluxes of the slurry with 30% particle mass fraction and that of pure HFE7100 were 59 and 69 W/cm^2, respectively. **Figure 12** showed the heat transfer coefficients of slurries with several particle mass fractions vs. the heater surface temperature at a flow rate of 7.15 cc/s. The figure showed the heat transfer coefficients of nano-PCM slurries peak at 60–63°C.

The reported heat transfer results in **Figures 10–12** were the average values from three consecutive repeated tests. Depending on the temperature of the heater, a heat loss of 4.5% of the electrical power input was estimated by calibration. The heat flux at the surface of the copper plate was obtained from the measured electrical power. Data in **Figures 10 and 11** were the heat transfer coefficients from five consecutive tests after accounting for the heat loss. All experiments had random

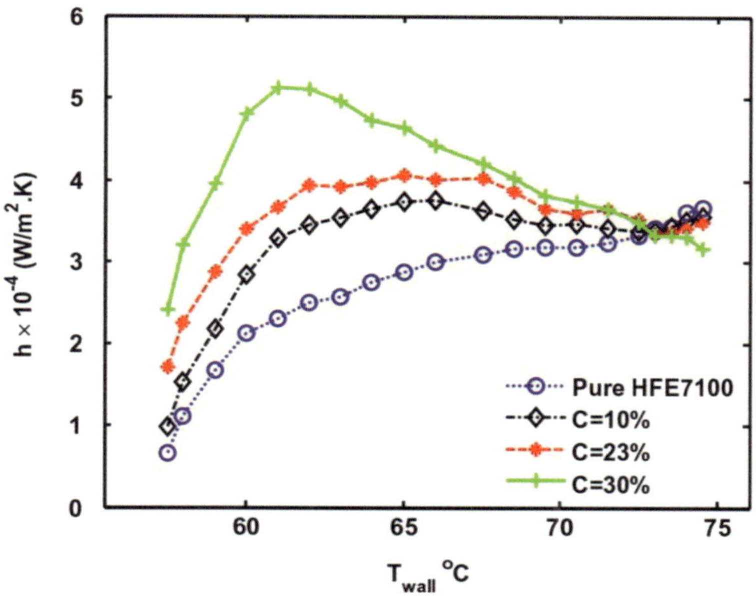

Figure 12.
Heat transfer coefficients of slurry with 30% particle mass fraction vs. heater surface temperature of full range (56–75°C) at the flow rate of 7.15 cc/s (note: the inlet T_{in} was controlled at 56°C, nozzle: TG0.76, H = 20 mm).

errors which occurred inevitably during measurements. These errors may be analyzed with the calculation of the root-mean-square of errors (RMSE) defined as RMSE = $\sqrt{\frac{1}{n}\left(\sum_1^n\left(\frac{h-h_{exp}}{h}\right)\right)}$ where n was the size of the sample (n = 5). The RMSE values for pure HFE7100, 10% slurry, 23% slurry, and 30% slurry are 0.013, 0.036, 0.04, and 0.049, respectively. During these measurements the maximum root-mean-square of errors (RMSE) was found to be less than 5% representing a reasonable degree of accuracy.

Our data from the slurries with nano-PCMs demonstrated very consistent thermal performance. The 30% nano-PCM slurry attains 97% of its initial heat transfer performance after 5000 thermal cycles. This implied that encapsulation of the nano-PCM particles with a shell made of materials such as silica, normally used to prevent coalescence of molten nanoparticles, was not needed for the slurry featuring HFE7100 and Field's alloy nano-PCM. The main reason was thought to be the existence of oxide shells around the nanoparticles. Oxidation of Field's alloy was unavoidable during the synthesis process and can provide a thin protective shell for the core material for a long time. Furthermore, two other possible mechanisms might be helping the bare nano-PCMs. First, the added perfluorooctyltriethoxysilane surfactant helped resist coalescence of molten Field's alloy nanoparticles and ensured the stability of colloidal suspension. Second, bare Field's alloy nanoparticles had residual charges which generate repulsive electrical force to prevent the agglomeration of molten nanoparticles.

5. Conclusions

In this chapter, a facile one-step method was developed for the production of Field's alloy nanoparticles or slurry using nanoemulsification technique. The composition, size, morphology, and thermal properties of as-prepared nanoparticles were characterized by XRF, TEM, and DSC, respectively. The slurry with modified Field's alloy nanoparticle dispersed in PAO or HFE7100 exhibits good thermal properties and stability, which showed promising potential applications in cooling of electronic device, engines, and other systems. Meanwhile, an experimental study was performed to investigate jet impingement heat transfer of Field's alloy nanoparticles-HFE7100 slurry. The Field's alloy nano-PCM absorbed heat during a phase change process from solid to liquid phase coupled with HFE7100 evaporation process. The study showed that the mass fraction of nanoparticles played an insignificant role in pressure drop but an important role on heat transfer performance. The high heat flux removal capability had been demonstrated by repeated closed loop test. Away from the critical heat flux, Field's alloy nano-PCM slurry provided a significant heat transfer enhancement due to the increase in the thermal capacity of the carrier fluid. Moreover, the nano-PCM slurries were able to maintain 97% of their heat removal capability after 5000 thermal cycles.

Acknowledgements

This work is supported by the by the National Natural Science Foundation of China (Grant No. 51672227) and the Fundamental Research Funds for the Central Universities (Grant No. YX2682015RC08, 2682017CY08 and 2682017CX089).

Conflict of interest

The authors declare no conflict of interest.

Nomenclature

A	area of the heat transfer (m^2)
c_p	heat capacity of the working fluid (J/kg K)
k	thermal conductivity of working fluid (W/m K)
h_{sl}	latent heat of the Field's alloy (J/kg)
h	heat transfer coefficient of jet impingement or spray (W/m^2 K)
Q	total power (W)
q	heat flux (W/m^2)
R	thermal resistance (K/W)
r	radius (m)
T	temperature (K)
T_f	temperature of inlet fluid (K)
ϕ	volume fraction of nanoparticles in the working fluid
Greeks	
q	density (kg/m^3)
s	melting time (s)
l	viscosity (Pa s)
Subscripts	
b	bulk
d	diameter of nozzle
f	carrier fluid
l	melted Field's alloy
m	melting
p	particle
w	wall
s	solid

Author details

Chaoming Wang[1,2]*, Xinran Zhang[1], Wenbing Jia[2], Wei Wu[3] and Louis Chow[3]

1 Applied Mechanics and Structure Safety Key Laboratory of Sichuan Province, School of Mechanics and Engineering, Southwest Jiaotong University, Chengdu, China

2 Key Laboratory for Advanced Technologies of Materials, Ministry of Education, School of Materials Science and Engineering, Southwest Jiaotong University, Chengdu, China

3 Department of Mechanical and Aerospace Engineering, University of Central Florida, Orlando, FL, USA

*Address all correspondence to: hbdxwcm@hotmail.com

References

[1] Kurtcebe C, Erim MZ. Heat transfer of a non-Newtonian viscoinelastic fluid in an axisymmetric channel with a porous wall for turbine cooling application. International Communications in Heat and Mass Transfer. 2002;**29**:971-982

[2] Liu M, Bruno F, Saman W. Thermal performance analysis of a flat slab phase change thermal storage unit with liquid-based heat transfer fluid for cooling applications. Solar Energy. 2011;**85**:3017-3027

[3] Forgber T, Radl S. Heat transfer rates in wall bounded shear flows near the jamming point accompanied by fluid-particle heat exchange. Powder Technology. 2017;**315**:182-193

[4] Hryniewicz P, Szeri AZ, Jahanmir S. Application of lubrication theory to fluid flow in grinding: Part I—Flow between smooth surfaces. Journal of Tribology-Transactions of the ASME. 2001;**123**:94-100

[5] Kim E. The influence of Reynolds number and heat transfer in fluid film lubrication. Mechanics Research Communications. 1996;**23**:441-448

[6] Wannapakhe S, Rittidech S, Bubphachot B, Watanabe O. Heat transfer rate of a closed-loop oscillating heat pipe with check valves using silver nanofluid as working fluid. Journal of Mechanical Science and Technology. 2009;**23**:1576-1582

[7] Navas J, Sanchez-Coronilla A, Martin EI, Teruel M, Gallardo JJ, Aguilar T, et al. On the enhancement of heat transfer fluid for concentrating solar power using Cu and Ni nanofluids: An experimental and molecular dynamics study. Nano Energy. 2016;**27**:213-224

[8] Yang JC, Li FC, He YR, Huang YM, Jiang BC. Experimental study on the characteristics of heat transfer and flow resistance in turbulent pipe flows of viscoelastic-fluid-based Cu nanofluid. International Journal of Heat and Mass Transfer. 2013;**62**:303-313

[9] Modak M, Sharma AK, Sahu SK. An experimental investigation on heat transfer enhancement in circular jet impingement on hot surfaces by using Al_2O_3/water nano-fluids and aqueous high-alcohol surfactant solution. Experimental Heat Transfer. 2018;**31**:275-296

[10] Rajendran DR, Sundaram EG, Jawahar P. Experimental studies on the thermal performance of a parabolic dish solar receiver with the heat transfer fluids SiC plus water nano fluid and water. Journal of Thermal Science. 2017;**26**:263-272

[11] Asadi A, Asadi M, Rezaniakolaei A, Rosendahl LA, Afrand M, Wongwises S. Heat transfer efficiency of Al_2O_3-MWCNT/thermal oil hybrid nanofluid as a cooling fluid in thermal and energy management applications: An experimental and theoretical investigation. International Journal of Heat and Mass Transfer. 2018;**117**:474-486

[12] Pal R. A novel method to determine the thermal conductivity of interfacial layers surrounding the nanoparticles of a nanofluid. Nanomaterials. 2014;**4**:844-855

[13] Song SH, Shen WD, Wang JL, Wang SC, Xu JF. Experimental study on laminar convective heat transfer of microencapsulated phase change material slurry using liquid metal with low melting point as carrying fluid. International Journal of Heat and Mass Transfer. 2014;**73**:21-28

[14] Tao YB, You Y, He YL. Lattice Boltzmann simulation on phase change heat transfer in metal foams/

paraffin composite phase change material. Applied Thermal Engineering. 2016;**93**:476-485

[15] Liu YS, Yang YZ. Preparation and thermal properties of $Na_2CO_3{\cdot}10H_2O$-$Na_2HPO_4{\cdot}12H_2O$ eutectic hydrate salt as a novel phase change material for energy storage. Applied Thermal Engineering. 2017;**112**:606-609

[16] Kahwaji S, Johnson MB, Kheirabadi AC, Groulx D, White MA. Fatty acids and related phase change materials for reliable thermal energy storage at moderate temperatures. Solar Energy Materials & Solar Cells. 2017;**167**:109-120

[17] Aydin AA, Aydin A. High-chain fatty acid esters of 1-hexadecanol for low temperature thermal energy storage with phase change materials. Solar Energy Materials & Solar Cells. 2012;**96**:93-100

[18] Deng YY, Yang LJ. Preparation and characterization of polyethylene glycol (PEG) hydrogel as shape-stabilized phase change material. Applied Thermal Engineering. 2017;**114**:1014-1017

[19] Lv PZ, Liu CZ, Rao ZH. Experiment study on the thermal properties of paraffin/kaolin thermal energy storage form-stable phase change materials. Applied Energy. 2016;**182**:475-487

[20] Chen Z, Feng S, Cao L, Fang G. Synthesis and thermal properties of shape-stabilized lauric acid/activated carbon composites as phase change materials for thermal energy storage. Solar Energy Materials & Solar Cells. 2012;**102**:131-136

[21] Han ZH, Cao FY, Yang B. Synthesis and thermal characterization of phase-changeable indium/polyalphaolefin nanofluids. Applied Physics Letters. 2008;**92**:2252

[22] Ma Y, Chu X, Tang G, Yao Y. Synthesis and thermal properties of acrylate-based polymer shell microcapsules with binary core as phase change materials. Materials Letters. 2013;**91**:133-135

[23] Hong Y, Ding S, Wu W, Hu J, Voevodin AA, Gschwender L, et al. Enhancing heat capacity of colloidal suspension using nanoscale encapsulated phase-change materials for heat transfer. ACS Applied Materials & Interfaces. 2010;**2**:1685-1691

[24] Wang Y, Xia Y. Bottom-up and top-down approaches to the synthesis of monodispersed spherical colloids of low melting-point metals. Nano Letters. 2004;**4**:2047-2050

[25] Fang Y, Kuang S, Gao X, Zhang Z. Preparation and characterization of novel nanoencapsulated phase change materials. Energy Conversion and Management. 2008;**49**:3704-3707

[26] Tiarks F, Landfester K, Antonietti M. Preparation of polymeric nanocapsules by miniemulsion polymerization. Langmuir. 2001;**17**:908-918

[27] Canter N. Heat transfer fluid based on nanoparticles dispersed in ionic liquids. Tribology and Lubrication Technology. 2014;**70**:10-11

[28] Manikandan S, Rajan KS. New hybrid nanofluid containing encapsulated paraffin wax and sand nanoparticles in propylene glycol-water mixture: Potential heat transfer fluid for energy management. Energy Conversion and Management. 2017;**137**:74-85

[29] Krishnamurthy MR, Kumar G, Gireesha BJ, Rudraswamy NG. MHD flow and heat transfer (PST and PHF) of dusty fluid suspended with alumina nanoparticles over a stretching sheet embedded in a porous medium under the influence of thermal radiation. Journal of Nanofluids. 2018;**7**:527-535

[30] Vu T, Tran TN, Xu JJ. Single-phase flow and heat transfer characteristics of ethanol/polyalphaolefin nanoemulsion fluids in circular minichannels. International Journal of Heat and Mass Transfer. 2017;**113**:324-331

[31] Yu LY, Liu D, Botz F. Laminar convective heat transfer of alumina-polyalphaolefin nanofluids containing spherical and non-spherical nanoparticles. Experimental Thermal and Fluid Science. 2012;**37**:72-83

[32] Wang X, Niu J, Li Y, Wang X, Chen B, Zeng R, et al. Flow and heat transfer behaviors of phase change material slurries in a horizontal circular tube. International Journal of Heat and Mass Transfer. 2007;**50**:2480-2491

[33] Yamagishi Y, Takeuchi H, Pyatenko AT, Kayukawa N. Characteristics of microencapsulated PCM slurry as a heat-transfer fluid. AICHE Journal. 2010;**45**:696-707

[34] Zeng R, Wang X, Chen B, Zhang Y, Niu J, Wang X, et al. Heat transfer characteristics of microencapsulated phase change material slurry in laminar flow under constant heat flux. Applied Energy. 2009;**86**:2661-2670

[35] Nan CW, Birringer R, Clarke DR, Gleiter H. Effective thermal conductivity of particulate composites with interfacial thermal resistance. Journal of Applied Physics. 1997;**81**:6692-6699

[36] Hu X, Zhang Y. Novel insight and numerical analysis of convective heat transfer enhancement with microencapsulated phase change material slurries: Laminar flow in a circular tube with constant heat flux. International Journal of Heat and Mass Transfer. 2002;**45**:3163-3172

[37] Wang C, Zhang X, Su M. Synthesis and thermal stability of Field's alloy nanoparticles and nanofluid. Materials Letters. 2017;**205**:6-9

[38] Wu W, Chow LC, Wang CM, Su M, Kizito JP. Jet impingement heat transfer using a Field's alloy nanoparticle – HFE7100 slurry. International Journal of Heat and Mass Transfer. 2014;**68**:357-365

[39] Vehkamaki H. Classical Nucleation Theory in Multicomponent Systems. New York: Springer; 2006

[40] Incropera FP, Dewitt DP. Introduction to Heat Transfer. John Wiley & Sons; 2006

[41] Maxwell JC. A Treatise on Electricity and Magnetism. New York: Dover; 1954

[42] Kizilkan O, Dincer I. Borehole thermal energy storage system for heating applications: Thermodynamic performance assessment. Energy Conversion and Management. 2015;**90**:53-61

[43] Miro L, Oro E, Boer D, Cabeza LF. Embodied energy in thermal energy storage (TES) systems for high temperature applications. Applied Energy. 2015;**137**:793-799

[44] Fox RW, McDonald A. Introduction to Fluid Mechanics. New Jersey: John Wiley & Sons; 2011

[45] Martin H. Advances in Heat Transfer. Vol. 13. New York: Academic Press; 1977